T0192312

# Structured Glass-Fiber Catalysts

# Structured Glass-Fiber Catalysts

Authored by
Andrey Zagoruiko and Sergey Lopatin

CRC Press is an imprint of the
Taylor & Francis Group, an informa business

CRC Press
Taylor & Francis Group
6000 Broken Sound Parkway NW, Suite 300
Boca Raton, FL 33487–2742

First issued in paperback 2021



CRC Press is an imprint of Taylor & Francis Group, an Informa business



Printed on acid-free paper

ISBN-13: 978-0-367-25385-1 (hbk)
ISBN-13: 978-1-03-208426-8 (pbk)



**Library of Congress Cataloging-in-Publication Data**

Names: Zagoruiko, Andrey, author. | Lopatin, Sergey I., author. Title: Structured glass-fiber catalysts / authored by Andrey Zagoruiko and Sergey Lopatin.
Description: First edition. | Boca Raton, FL : CRC Press/Taylor & Francis Group, 2020. | Includes bibliographical references and index.
Identifiers: LCCN 2019033397 | ISBN 9780367253851 (hardback ; acid-free paper) | ISBN 9780429317569 (ebook)
Subjects: LCSH: Catalysts—Materials. | Glass fibers.
Classification: LCC TP159.C3 Z34 2020 | DDC 660/.2995—dc23
LC record available at https://lccn.loc.gov/2019033397

**Visit the Taylor & Francis Web site at**
**www.taylorandfrancis.com**

**and the CRC Press Web site at**
**www.crcpress.com**

# Dedication

*Andrey Zagoruiko with love and gratitude devotes this work to*

*his wife Rumyana,*

*his parents Nickolay and Valentina,*

*his sister Galina,*

*his children Maria, Nickolay, Dmitry and Ksenia.*

*Sergey Lopatin with love and gratitude devotes this work to*

*his wife Olga,*

*his father Alexey,*

*his children Olga and Alexey.*

# Contents

# Preface

The development of novel catalytic processes or efficiency improvements of existing ones is an important task, having a high practical value in such areas as chemistry, petrochemistry, oil processing, environmental protection, and in particular in the abatement of toxic atmospheric pollutants from automobile transport and industrial and energy manufacturing facilities.

One of the most promising methods for the solution of this task is the development of processes on the base of catalysts with a new geometric shape, which are characterized by higher efficiency of heat and mass transfer, thus providing for the maximum use of the catalytic potential of the catalysts' active components.

The breakthrough in this area can be achieved owing to the application of novel catalysts on the base of glass-fiber supports. The most important distinction between glass-fiber catalysts (GFCs) and traditional catalysts is their specific geometric structure, high mechanical strength, and flexibility, opening up unique possibilities for the creation of new types of differently shaped structured beds with reduced pressure drop and improved heat and mass transfer. This, in turn, makes it possible to create new designs of catalytic reactors.

At the same time, the specificity of such catalysts raises certain problems with regard to their practical application. In particular, until now, the various packings of microfibrous catalysts have not yet been studied fully enough in regard to the quantitative description of heat and mass transfer processes, and it significantly complicates the scale up and optimization of GFC-based reactors. In many cases, we may face practical difficulties in incorporating GFCs into existing reactors, which were designed to use the conventional catalysts' shapes (pellets or monoliths).

The current text is dedicated to the issues of the synthesis and investigation of GFCs, as well as to development of the engineering basis for the commercial catalytic processes applying these catalysts.

# Acknowledgments

The authors consider it their honor to express their gratitude to all their colleagues and friends, without whom working on this paper would be impossible:

Dr. Pavel Mikenin, Danil Pisarev, Dmitry Baranov, Sergey Zazhigalov, Dr. Sergey Vanag, Prof. Eugene Paukshtis, Dr. Tatyana Larina, Dr. Svetlana Cherepanova, Dr. Svetlana Yashnik, Ilya Pakharukov, Dr. Oleg Klenov, Dr. Sergey Veniaminov, Dr. Arkadii Ischenko, Nina Rudina, Dr. Nikolay Yazykov, Ksenia Golyashova, Prof. Vadim Yakovlev (Boreskov Institute of Catalysis), Prof. Pavel Tsyrulnikov, Dr. (Institute for Hydrocarbon Processing), Dr. Andrey Elyshev (Tyumen State University), Yulia Kotolevich, Dr. Maxim Popov, Dr. Alexander Bannov, Ulyana Ovchinnikova (Novosibirsk State Technical University), Dr. Alexander Zykov and Sergey Anichkov (All-Russian Thermal Engineering Institute), who directly took part in the described research.

Prof. Valentin Parmon, Prof. Valerii Kirillov, Dr. Alexander Kulikov, Prof. Gilbert Froment, Ilya Zolotarskii, Prof. Yurii Matros, Dr Liudmila Bobrova, Dr. Nadezhda Vernikovskaya, Dr. Grigorii Bunimovich, Prof. Gianpiero Groppi, Prof. Dan luss, Prof. Anton Maximov, Vera Fadeeva, Igor Kim, Irina Kharina, Dr. Vassily Kruglyakov, Tamara Yudina, Liubov Poleschuk, Dr. Alexey Suknev, Dr. Viktor Chumachenko, Dr. Alyona Ovchinnikova, Prof. Alexander Noskov, Dr. Olga Chub, Sergey Kildyashev, Prof. Sergey Reshetnikov, Valentina Shport—for useful discussions, advice, inspiration, and support.

Vladimir Serbinenko and all team at Sibtransservice Co.; Dr. Yurii Zhukov, Viktor Glotov, Vassily Yankilevich, Nikolay Proskokov (Byisk Oleum Plant); Alexander Pateev (Moscow Metro); Dr. Alexander Beskopylnii and Andrey Kovalenko (Volgograd department of Boreskov Institute of Catalysis); Nayil Gilmutdinov and Ilziya Nazmieva (Nizhnekamskneftehim Co.)—for active support in the arrangement and performance of the pilot and semi-industrial tests of GFCs.

We also address our special thanks to:

Eugenia Bobatkova, who made a cover design and many excellent photographs for this book; Olga Lopatina for the great editorial work of this book.

# Acknowledgments

The authors consider it their honor to express their gratitude to all their colleagues and friends, without whom working out this paper would be impossible.

Dr. [illegible], [illegible], Dmitry Baranov(?), Sergey [illegible], Dr. [illegible], Prof. Eugene [illegible], Dr. Tatyana [illegible], Dr. Svetlana Cherepanova(?), Dr. Svetlana [illegible], [illegible], Dr. Oleg [illegible], Dr. Sergey [illegible], Dr. [illegible], [illegible] Radina, Dr. [illegible], [illegible], Prof. Vasily Yakovlev (Boreskov Institute of [illegible]), Prof. Pavel [illegible], Dr. (Institute of Hydrocarbon Processing), Dr. Andrey [illegible], [illegible], Dr. Maxim Popov(?), Dr. Alexander [illegible], [illegible] Novosibirsk State Technical University, Dr. Alexander [illegible] and Sergey [illegible] (All-Russian Thermal Engineering Institute) [illegible] directly to the development research.

Dr. [illegible], Prof. Valery [illegible], Dr. Alexander [illegible], Professor of [illegible], Prof. [illegible], Dr. [illegible], [illegible], Dr. [illegible], Prof. [illegible], [illegible], [illegible], Dr. Alexey [illegible], [illegible], Dr. Alexander [illegible], [illegible], Prof. [illegible] for [illegible] discussions, inspiration, and support.

Vladimir [illegible] [illegible] Volgograd [illegible] Department of [illegible] Institute of Chemical Physics [illegible] for active support in the organization and performance of the pilot and commercial tests of [illegible].

We also would like to especially thank [illegible].

[illegible], who made the cover design in my excellent [illegible] the book. Oleg [illegible] for the organizational work of this book.

# About the Authors

**Andrey Zagoruiko** Doctor of Technical Science (rehab), leading researcher in the Boreskov Institute of Catalysis (Novosibirsk, Russia). Graduated from Moscow Institute of Fine Chemical Technology. More than 30 years of experience in the area of mathematical modeling and engineering of catalytic processes, as well as development of new catalytic technologies for oil and gas processing, petrochemistry, and environmental protection. Author of more than 300 scientific publications, including over 70 patents. Lecturer in Novosibirsk State University for more than 20 years. A permanent member of the scientific committees of the highly ranked international ChemReactor and HydroCat conferences. Guest editor for *Chemical Engineering Journal*, *Chemical Engineering and Processing: Process Intensification*, and *Catalysis Today* journals. Member of the editorial board of the *Reviews in Chemical Engineering* and *Catalysis in Industry* journals. Laureate of the Russian Academy of Sciences Award in the name of Valentin Koptyug, gold medalist of the Russian National Exhibition Center and the Siberian Fair for the developments in the area of environmental protection technologies.

**Sergey Lopatin** Engineering Group Head in Boreskov Institute of Catalysis (Novosibirsk, Russia). Graduated from Kuzbass Polytechnic Institute. Expert in the area of mass transfer in catalytic structures, engineering of catalytic processes, development and modeling of catalytic reactors, development of new catalyst shapes and technologies for their manufacturing. Fourteen years of research experience. Scientific and practical interests include catalytic processes on the base of microfibrous catalysts for environmental protection, waste recycling, and environmentally friendly combustion of fuel. Author of more than 80 scientific publications, including over 30 patents.

# 1 Current State of Research and Development in the Field of Catalysts With Various Shapes and Their Practical Application

## 1.1. INTRODUCTION

Today, catalysts and catalytic technologies are at the root of the modern chemical industry, oil and gas processing, and the petrochemical industry that produce critically important products such as basic chemicals, fuel, polymers, fertilizers, etc. A variety of technologies for environmental protection and utilization of wastes are also based on catalysts. Catalytic technologies play an increasing role in a new green energy, as well as in the field of renewable feedstocks and fuel. Nowadays more than 90% of industrial products are manufactured with the use of catalysts; moreover, they are involved in the production of the most advanced products with the high level and depth of feedstock processing. The further development of energy production and energy storage sectors, as well as novel industry in the direction of substantial increase of feedstock, energy, and environmental efficiency is impossible without advanced growth of catalytic processes.

The main progress drivers in catalysis can be identified as follows:

- Chemical (in case a new catalytic technology is based on a new reaction or new reaction routes for target products production)
- Catalytic (based on application of a new catalyst in a known reaction)
- Engineering (based on application of a new engineering approach to performance of a known catalytic reaction)

Engineering approaches to create a new catalytic technology are based on the following problems solutions:

- Development of new catalyst shapes and structured catalyst beds (pellets, monoliths, foams, fabrics, etc.)
- Development of new and optimization of existing configurations of catalytic beds inside reactors (fixed beds, tubular reactors, fluidized and moving beds, etc.)
- Development of reactors with new methods of energy management (e.g., induction heated reactors)

- Development of catalytic technologies and reactors, combining reaction and product separation
- Development of new methods of mass and heat exchange processes in catalytic reactors
- Application of dynamic regimes of reaction performance (unsteady-state and sorption-enhanced catalytic processes)

The first and second problems noted are of special interest within the context of the given work.

## 1.1. TRADITIONAL CATALYST SHAPES

The solid heterogeneous catalysts are the most widely used type of the catalytic systems in practice. To reach the catalytic potential of their active component, the latter is usually supported at the surface of various supports with the developed internal porous structure and high specific surface area to provide high catalytic efficiency of the material. The selection of the support is commonly dictated by the specific requirements of the target reaction and its performance conditions. Currently, supports on the base of alumina, silica, activated carbon, and titania prevail in industrial applications.

From an engineering point of view the selection of the optimum catalyst support is usually focused on the achievement of the maximum stimulation of heat and mass transfer processes. In addition, the requirements for mechanical strength, thermal stability (both in terms of long-term operation stability at elevated temperatures and resistance to thermal shock), and hydraulic resistance to moving reaction fluid (or pressure drop) are often applied in the support selection procedure.

Intensification of heat and mass exchange processes can be reached by proper selection and optimization of the shape of the catalyst particles. Technically, the minimization of diffusion resistance requires the smallest possible catalyst particles; however, in reality the highly dispersed powder catalysts have rather limited application (e.g., fluidized beds, slurry beds), and their use is difficult or even impossible due to technological problems with the arrangement of catalyst beds and their high pressure drop. Different types of granulated and structured catalysts are applied to overcome the difficulties.

## 1.2. FIXED BEDS OF GRANULAR CATALYSTS

Granular catalysts[1] are the most traditional and widely used type of catalytic system. They usually they have a shape of cylinders, balls, rings, and sometimes more complicated shapes (see Figure 1.1).

The typical size of catalyst pellets can vary for different catalytic processes, but in general it ranges from a few tenths of millimeters up to a few centimeters, with the most typical range being from 2 to 10 mm. The lower limit in this range is usually defined by pressure drop requirements, the upper one by requirements on internal and external diffusional resistance.

Cylinder and spherical pellets are the easiest to produce; however, they are characterized by a rather high pressure drop, as well as the relatively low level of the external specific surface and noticeable diffusional resistance in fast catalytic reactions.

**FIGURE 1.1** Typical shapes of catalyst pellets.

*Source:* Photo collage by Eugenia Bobatkova.

These problems can be minimized by the application of the ring-shaped pellets, as well as pellets of more complicated shapes[2,3] (see Figure 1.1). The fixed beds with such pellets have higher void fraction (from 0.5 to 0.7 instead of ~0.4 for the beds of spherical and cylindrical pellets); this results in significant decrease of bed pressure drop. They are also characterized by a higher external specific surface and a lower value of equivalent hydrodynamic diameter (with the same external size of the pellets), and this allows a significant decrease of diffusion resistance. The drawbacks of such pellets are the high technical complexity of their production and their lower thermal stability, especially in cases of complex shapes (rings with internal bridges, multilobes) and thermal shocks.

From the hydrodynamic properties point of view, the fixed beds of granular catalysts have a good ability to turbulize the reaction fluids, even at low fluid velocities, thereby providing a high intensity of heat and mass transfer.[4] These beds are characterized with full hydrodynamic isotropy, that is, the equality of hydraulic resistance coefficients for fluid movement in any direction inside the bed. The last property, along with the relatively high bed pressure drop, provides an excellent leveling ability of fixed granular beds, making it possible to minimize the non-uniformities of the flow distribution across the bed volume.

On the other hand, the high pressure drop of the fixed granular beds may become a significant disadvantage in catalytic reactors with high reaction fluid velocities.

## 1.3. MONOLITH CATALYSTS

A significant decrease of the catalyst pressure drop can be achieved by the application of the structured catalysts, also known as monolith catalysts or honeycomb blocks.[5–8] Such catalysts represent themselves as blocks with a regular system of straight parallel internal channels (see Figure 1.2).

**FIGURE 1.2** Monolith catalysts.

*Source:* Photo collage by Eugenia Bobatkova.

The size of the channels usually varies in the range of a few tenths of a millimeter up to a few millimeters; the cross-section of a channel can be square, rectangle, triangle, polygon, or round. Minimum channel size is defined by pressure drop requirements and manufacturing complexity; the maximum size is limited by diffusion resistance requirements. Externally the monolith may have a shape of a cylinder or prism with a square, rectangle, or polygon cross-section. The external shape and size of catalytic blocks are usually determined by the technological simplicity of their production and application.

The base of monolith catalyst is typically made of thermo-stable ceramics, though metal and carbon supports are also known.

The main advantage of monolith honeycomb catalysts is their extremely low pressure drop, which is important for the performance of the catalytic reactions under high velocities of fluid movements, for example, during purification of exhaust gases in combustion engines or flue gases at thermal electric power plants.

From the internal fluid dynamics point of view, monolith blocks with parallel channels have an absolute spatial anisotropy—the fluid can move inside them only along the channels; it is impossible to move in any other direction. It makes any redistribution of the flows between the channels impossible; therefore, the ability to level the non-uniformities of the initial flow distribution across the inlet section is zero. This drawback can be reduced by the application of the monoliths with inter-channel fluid passages.[9] However, such monoliths are hard to produce and are expensive; hence, their practical application is very limited.

Another fluid dynamic problem is the laminar nature of the fluid flows inside the channels even at high fluid velocities.[10] This results in a low efficiency of heat and mass transfer processes inside the blocks.

Ceramic catalytic monoliths are manufactured by extrusion of the support or catalytic material with a further possible application of a secondary support and/or an active component, in all cases followed by thermal treatment. Not every catalytic material can be successfully extruded, and not every support with the applied active component can successfully withstand the thermal treatment. This significantly limits the range of the supports and catalytic materials that can be used in the production of monolith catalysts and the variety of external shapes of the producing monoliths. Moreover, the ceramic monoliths may have a low mechanical strength and a low resistance to thermal shocks, which is even more important.

## 1.4. CATALYSTS WITH FOAM SUPPORTS

The relatively new types of catalysts use metal or ceramic foams as supports.[11–15] Such supports are characterized with a regular geometric structure with a well-controlled and uniform size of internal channels (see Figure 1.3).

The regular structure of the foam supports ensures their low pressure drop, as well as possibility of effective redistribution of the reaction flow, making it possible to improve the uniformity of flow distribution along the catalyst volume. The possible drawbacks of the systems may include a laminar nature of the internal fluids in the channels and corresponding limited efficiency of heat and mass transfer.

**FIGURE 1.3** Structured catalysts on the base of foam supports.

*Source:* Photo collage by Eugenia Bobatkova.

## 1.5. CATALYSTS WITH FLEXIBLE METAL SUPPORTS

Attention is also paid to catalysts that use flexible metal ribbons and meshes as the supports. The combined use of flat and corrugated supports makes it possible to create structured catalytic blocks of various shapes[16–18] (see Figure 1.4).

The undoubted merit of the metal supports is their flexibility, resulting in high resistance to a mechanical impact (including vibration) and thermal shock. At the same time, they have the main advantage of all the structured blocks—low pressure drop. A high heat conductivity, which is important for catalytic reactions with significant heat effects, is also the benefit of the systems.

The supporting of significant amount of active components at the metal surface itself is impossible in many cases, so the surface is usually modified by the application of the external layer of the secondary support, which is then used for deposition of an active component.[19–23] This procedure is complicated by the different thermal expansion coefficients of the primary and secondary supports (e.g., steel and alumina). In this case, if the temperature changes during catalyst operation, it can lead to the destruction of the secondary support layer, resulting in catalyst

**FIGURE 1.4** Structured blocks on the base of flexible metal supports.

*Source:* Photo collage by Eugenia Bobatkova.

deactivation down to complete loss of activity. To some extent, this problem can be solved by the use of aluminum-containing alloys as a primary support; for example, in case of fechral alloy it is possible to create a firm film of alumina on the surface, which can be used as a secondary support or as a substrate for the secondary support.[24] However, the use of such supports is limited by their relatively low mechanical strength and low ability to bear the mechanical processing, such as corrugation.

The use of massive flat ribbons as a support, as with monolith catalysts, will result in the formation of laminar fluid with a relatively low efficiency of external mass transfer. Moreover, the flow redistribution between channels will be also missing, negatively influencing the uniformity of flow distribution in the block volume. These problems can be overcome by application of the supports in the form of permeable metal meshes, both flat and corrugated, as well as various structures made of a metal wire.[25–28] These structures, shown in Figure 1.5, have come to be known as wire-mesh monoliths.

**FIGURE 1.5** Structured blocks on the base of wire-mesh metal supports.

*Source:* Photo collage by Eugenia Bobatkova.

In general, the catalysts on the base of fibrous supports are promising for the development of novel catalytic structures, provided that the problems with making a mechanically strong and thermally stable active component coverage at the surface will be successfully solved.

## REFERENCES

1. Boreskov G.K. 1986. *Heterogeneous catalysis*. Novosibirsk: Nauka.
2. Afandizadeh S., Foumeny E.A. 2001. Design of packed bed reactors: Guides to catalyst shape, size, and loading selection. *Appl Therm Eng*. 21: 669–82.
3. Mariani N.J., Mocciaro C., Keegan S.D. et al. 2009. Evaluating the effectiveness factor from a 1D approximation fitted at high Thiele modulus: Spanning commercial pellet shapes with linear kinetics. *Chem Eng Sci*. 64: 2762–6.
4. Aerov M.E., Todes O.M., Narinskii D.A. 1979. *Apparatuses with stationary granular bed: Hydraulic and heat basis of operation*. Leningrad: Chemistry.
5. Birtigh G., Parbel H., Sauer H. 1975. US Patent No.3926565. Apparatus for cleaning exhaust gases.
6. Ridgway S.L. 1965. US Patent No.3211534. Exhaust control.
7. Boger T., Heibel A.K., Sorensen C.M. 2004. Monolithic catalysts for the chemical industry. *Ind Eng Chem Res*. 43: 4602–11.
8. Mochida Shigeru, Kojima Masaru, Hijikata Toshihiko. 1983. US Patent No.4396664. Ceramic honeycomb structural body.
9. Gulati S.T. 1977. US Patent No.4042738. Honeycomb structure with high thermal shock resistance.

10. Klenov O.P., Pokrovskaya S.A., Chumakova N.A. et al. 2009. Effect of mass transfer on the reaction rate in a monolithic catalyst with porous walls. *Catal Today*. 144: 258–64.
11. Pestryakov A.N., Fyodorov A.A., Shurov V.A. 1994. Foam metal catalysts with intermediate support for deep oxidation of hydrocarbons. *React Kinet Catal L*. 53: 347–52.
12. Richardson J.T., Remue D., Hung J-K. 2003. Properties of ceramic foam catalyst supports: Mass and heat transfer. *Appl Catal A-Gen*. 250: 319–29.
13. Giani L., Groppi G., Tronconi E. 2005. Mass-transfer characterization of metallic foams as supports for structured catalysts. *Ind Eng Chem Res*. 44: 4993–2.
14. Twigg M.V., Richardson J.T. 2007. Fundamentals and applications of structured ceramic foam catalysts. *Ind Eng Chem Res*. 46: 4166–77.
15. Incera Garrido G., Patcas F.C., Lang S. et al. 2008. Mass transfer and pressure drop in ceramic foams: A description for different pore sizes and porosities. *Chem Eng Sci*. 63: 5202–17.
16. Fischer F.C. 1933. US Patent No.1932927. Device for converting carbon monoxide.
17. Kirillov V.A., Bobrin A.S., Kuzin N.A. et al. 2004. Compact radial reactor with a structured porous metal catalyst for the conversion of natural gas to synthesis gas: Experiment and modeling. *Ind Eng Chem Res*. 43(16): 4721–31.
18. Sanz O., Echave F.J., Romero-Sarria F. et al. 2013. Advances in structured and microstructured catalytic reactors for hydrogen production. *Renewable hydrogen technologies*/ed. by L.M. Gandía, G. Arzamendi, P.M. Diéguez. Amsterdam: Elsevier. Chapter 9: 201–24.
19. Retallick W.B. 1988. US Patent No.4762567. Washcoat for a catalyst support.
20. Fukuhara C., Hyodo R., Yamamoto K. et al. 2013. A novel nickel-based catalyst for methane dry reforming: A metal honeycomb-type catalyst prepared by sol—gel method and electroless plating. *Appl Catal A-Gen*. 468: 18–25.
21. Meille V. 2006. Review on methods to deposit catalysts on structured surfaces. *Appl Catal A-Gen*. 315: 1–17.
22. Pacheco Benito S., Lefferts L. 2010. The production of a homogeneous and well-attached layer of carbon nanofibers on metal foils. *Carbon*. 48(10): 2862–72.
23. Seijger G.B.F., Palmaro S.G., Krishna K. et al. 2002. In situ preparation of ferrierite coatings on structured metal supports. *Micropor Mesopor Mat*. 56(1): 33–45.
24. Sigaeva S., Temerev V.L., Borisov V.A. et al. 2015. Pyrolysis of methane on fechral resistive catalyst with additions of hydrogen or oxygen to the reaction mixture. *Catalysis in Industry*. 7(3): 171–4.
25. Jiang Z., Chung K-S., Kim G-R. et al. 2003. Mass transfer characteristics of wire-mesh honeycomb reactors. *Chem Eng Sci*. 58: 1103–11.
26. Sun H., Shu Y., Quan X. et al. 2010. Experimental and modeling study of selective catalytic reduction of NOx with $NH_3$ over wire mesh honeycomb catalysts. *Chem Eng J*. 165: 769–75.
27. Banús E.D., Sanz O., Milt V.G. et al. 2014. Development of a stacked wire-mesh structure for diesel soot combustion. *Chem Eng J*. 246: 353–65.
28. Porsin A.V., Kulikov A.V., Rogozhnikov V.N. et al. 2015. Catalytic reactor with metal gauze catalysts for combustion of liquid fuel. *Chem Eng J*. 282: 233–40.

# 2 Glass-Fiber Catalysts

## 2.1. GENERAL DESCRIPTION

Glass-fiber catalysts (GFCs) are a revolutionary type of catalytic system that utilize glass microfibers as structured supports (see Figure 2.1). Active components for these systems are usually selected from a wide range of various metals (Pt, Pd, Rh, Ir, Ag, Au, Fe, Cr, Co, N, Mn, Pb, Cu, etc.) and/or their oxides, depending on the specifics of the target application.

These catalysts have been known about for a rather long time,[1–9] but active development and research on GFCs started only in 1990s in the Institute of Chemical Physics of the Russian Academy of Sciences under the guidance of Professor V. V. Barelko.[10–13] Since the late 1990s, systematic studies of GFCs, mostly based on the use of the noble metals (Pt, Pd), have been started at the Boreskov Institute of Catalysis.[14–30]

It has been shown that GFCs demonstrate high activity and selectivity toward the target products in a number of reactions; moreover, they also show high resistance to deactivation in various aggressive media.[20,27–34] Their major advantage is an extra-low (below 0.1% mass) content of the noble metals, ensuring relatively low cost.

Later, investigations of GFCs were also started at Saint Petersburg University of Technology and Design in Russia,[35,36] the Technion Institute of Technology in Haifa, Israel,[37] the Institute for Hydrocarbon Processing in Omsk, Russia,[38] the Technische Universität, Dresden, Germany,[39] the North University of China, Taiyuan, China,[40] and many other universities and R&D companies worldwide.

Currently, GFCs are considered to be promising catalysts for the numerous reactions, including the following, as well as many others:

- Deep oxidation of CO, hydrocarbons, and organic substances[26–32,35–38,40–46]
- Deep oxidation of halogen-organic substances[20,34]
- Wastewater denitrification[47,48]
- Oxidation of $SO_2$[19,24,29,33,49–53]
- Selective oxidation of $H_2S$[54–59]
- Production of synthesis gas[60]
- Hydrosilation[61]
- Low-temperature reduction of $NO_x$[62]
- Methane chlorination[63]
- Oxychlorination of light olefins[64]
- Gas-phase biocatalysis[65]
- Photocatalysis[66–70]
- Selective hydrogenation[71–75]

**FIGURE 2.1** Glass-fiber catalyst.

*Source:* Photo collage by Eugenia Bobatkova.

## 2.2. SYNTHESIS OF GLASS-FIBER CATALYSTS

The simplest GFC synthesis method is based on an incipient wetness impregnation of the initial glass-fiber materials (fabrics, glass wool, or glass pads) with the solution of active component precursor, usually followed by drying and thermal treatment.[11] Basically, this approach is similar to the synthesis method of conventional catalysts with traditional supports.

Preliminary procedures such as washing and incineration of the support for the removal of dirt, dust, and organic lubricants from the surface of glass fibers can be performed before the impregnation.

Pre-treatment may also include leaching. In some cases, leaching is required to remove the alkaline compounds from the glass surface to improve further impregnation.[3,6] In other cases, leaching is necessary to make the local density non-uniformities in the glass volume.[29,76] Although the active metals (Pt or Pd) may be stabilized in the low-density areas of the glass matrix in the form of highly dispersed clusters with a size of ~1 nm, it is believed that such clusters are located inside the glass volume, in the subsurface layers of the glass fiber structure.[75]

The production method for these catalysts is rather complicated and may result in the formation of liquid wastes and loss of the precious metals.

The surface thermal synthesis (STS) method is a promising alternative for GFC production. It is based on formation of the catalytically active component directly at the glass-fiber surface during the thermal treatment of the supported combined precursor, containing the active metal compound and combustible fuel additive. The interaction of the fuel additive with oxygen produces the highly dispersed and potentially highly active metal or metal oxide surface particles under fast heating. This method has been applied for a rather long time to synthetize the catalysts with the conventional supports.[77–85] Over the last decade significant progress has been achieved in STS application for manufacturing of GFCs.[86–91]

The required activity of GFCs based on the noble metals can be achieved at the relatively low amount of active metal (< 0.1% mass). Catalysts using the oxides of the transient metals (which are less active than noble metals in most cases) as the active components usually require much higher content of the active component. However, a typical specific surface area of the glass fiber fabrics is not greater than 1–3 $m^2/g$, thus limiting the quantity of the supported oxide at the level of 0.7–0.8%, which is not enough for the majority of applications. Moreover, the strength of oxide binding to the glass surface can be too low for the practical catalyst handling and application.

The surface of the glass microfibers can be modified by deposition of the external layer of porous secondary support[4,55,56,92,93] (e.g., $SiO_2$) to increase the mass content of the active component. It allows for the increase of the specific surface area of the glass-fiber support up to 20–30 $m^2/g$. The most appropriate material for this layer is a porous silica, as it has a good adhesion to the glass surface, thus simplifying the supporting procedure. Additionally, the coefficient of silica thermal expansion is almost equal to that of the glass, providing a high resistance of such modified support to high temperature and thermal shocks. Except for $SiO_2$, the secondary support can be made of alumina, porous carbon,[94] or titania.[95] Such modified support can then be used for the synthesis of various GFCs, based on the deposition of almost any individual oxide or combination of oxides. It expands the area of potential GFC[44,55–57,59,96] application significantly.

The selection of a proper glass support is an important issue in GFC manufacturing. Conventional silica and boron-silica glasses have insufficient thermal stability for many practical applications. High thermal stability can be reached by the use of a special high-silica glass promoted by zirconia,[29] but this glass has a limited market availability and it is too expensive for economically reasonable production costs of GFCs.

It is much more promising to use the thermo-stabilized high-silica glasses, such as KT-11-TO fabric produced in Russia.[97] This fabric contains 94.5–96.0% of $SiO_2$, up to 3.5% $Al_2O_3$, up to 1% CoO, and up to 1% $SO_3$, and after the thermal pretreatment at 1000°C it is characterized by the presence of a weak absorption band, which has a wavenumber of $\nu$ = 3620–3650 $cm^{-1}$ and a half-width of more than 150 $cm^{-1}$ in the infrared spectrum; the value of the specific surface area, measured by alkaline titration method, is $S_{Na} = 3\ m^2/g$ and $S_{Na}/S_{Ar} = 3$. A low-intensity hydroxyl absorption band in the IR spectrum and its long half-width (above 100 $cm^{-1}$), as well as a low $S_{Na}/S_{Ar}$ ratio (less than 5), correspond to a small amount of hydroxyl groups in the volume of the glass microfibers. The absence or negligible content of these groups in the glass structure makes impossible the dehydroxylation of the support, and therefore provides its high thermal stability (up to 1100°C).

The order of the catalyst manufacturing stages can also differ. Most of the described methods are based on the use of the initially prepared glass-fiber fabric (GFF), but it is also possible to use the impregnation, thermal processing, and other required technological steps of GFC synthesis with single glass-fiber threads at first, followed by arrangement of these threads into the fabric.[98]

## 2.3. PLATINUM CATALYST IC-12-S111

The STS method[90] was applied to the synthesis of a Pt-containing GFC for reactions involving deep oxidation of CO, hydrocarbons, and organic compounds, and this GFC later received the IC-12-S111[46] trademark.

Platinum was selected as the active component due to its high activity in the target reactions. The available Pt precursors—hexachloroplatinum acid and dichloride of tetraammonium Pt—were used in the synthesis of GFC samples.

More than 60 different Pt-GFC samples were synthesized with variation of:

- Glass-fiber support type
- Nature and content of Pt precursor
- Nature and content of fuel additives, secondary supports, and promotors
- Parameters of the thermal treatment

Ceria was considered as a potential promotor, and alumina as an additional secondary support.

The synthesized samples were tested in a model reaction of deep propane oxidation in the recycle flow installation. Reaction mixture composition in these experiments was as follows: 1% vol. $C_3H_8$, 20% vol. $O_2$, balance—nitrogen. The reactor heating was performed in the range of 50–450°C; the temperature rise rate was 4°C per minute. The flow rate of the reaction mixture was 1000 mL/min.

It was found that catalyst activity increased when the Pt content was increased in the range of 0.07–0.1% mass, but further amounts of Pt did not cause the activity to increase.

Figure 2.2 shows the experimental result for various GFC samples with Pt content of ~0.1% mass. It is seen that the activity of the GFC produced from dichloride of Pt tetraammonium is higher than for the sample produced from hexachloroplatinum acid. However, as the difference is negligible, both precursors are appropriate for GFC synthesis.

The catalysts with the additional support of alumina (up to 1% mass Al) and ceria (up to 1.5% mass Ce) have lower activity; the sample with both oxides demonstrates the worst performance among all the tested samples. Most probably, Al and

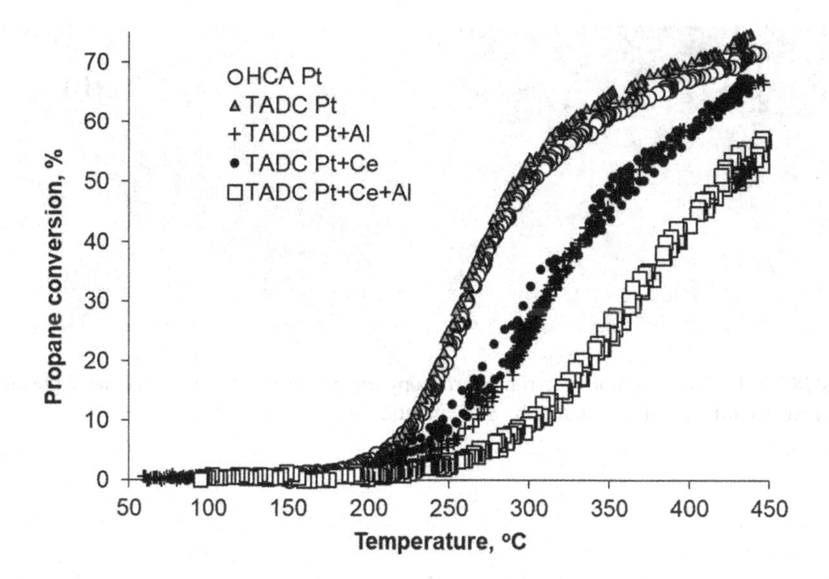


**FIGURE 2.2** Comparative activity of GFC samples in the reaction of deep oxidation of propane. HCA, hexachloroplatinum acid; TADC, tetraammonium dichloride.[91]

Ce oxides merely shield the active platinum species and, contrary to expectations, their presence produces no positive incentive effect.

It should be noted that the samples with alumina as the secondary support have a lower mechanical strength. Probably, in the course of the thermal treatment alumina interacts with the glass support; hence, alumina penetrates the glass bulk structure, causing glass destruction.

Finally, according to the test results, an optimum Pt/GFC sample should contain neither the oxide promotor nor the secondary support.

Figure 2.3 demonstrates the transmission electron microscopy (TEM) photographs of the initial GFF with supported precursor (Figure 2.3a) and GFC after the thermal treatment (Figure 2.3b). In both cases the interplanar distances can be attributed to Pt only according to the crystallographic database. No subsurface particles of the active component were observed. The size of the platinum particles in the initial GFF with supported precursor was 10–15 nm, decreasing to 2–4 nm in the GFC sample after the thermal processing.

Figure 2.4 shows the electron diffusion reflectance spectrum of the initial GFF with the supported precursor (curve 1) and Pt/GFC after the thermal treatment at 400°C (curve 2) and 700°C (curve 3).

The spectrum of the initial impregnated GFF contains typical bands of the precursor species (*d–d*-transition at ~36,000 $cm^{-1}$ and metal-ligand charge transfer band in platinum tetraammonium dichloride 46,000 $cm^{-1}$). Three types of platinum are observed in the GFC after the thermal treatment: ionic platinum $Pt^{2+}$ with maximum intensity in the range of 20,000 $cm^{-1}$, charged platinum clusters $Pt^{\delta+}$ (~38,000 $cm^{-1}$), and surface particles of metal Pt with the size of 5 nm and more (in the range above

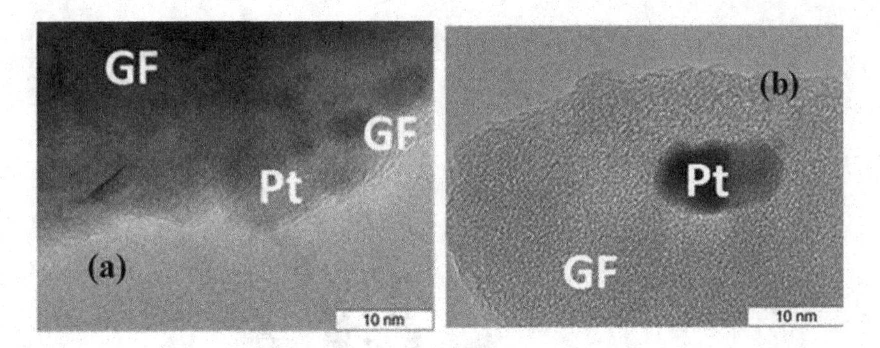


FIGURE 2.3 Transmission electron microscopy images of initial GFF with the supported Pt precursor (a) and GFC prepared by STS method (b).

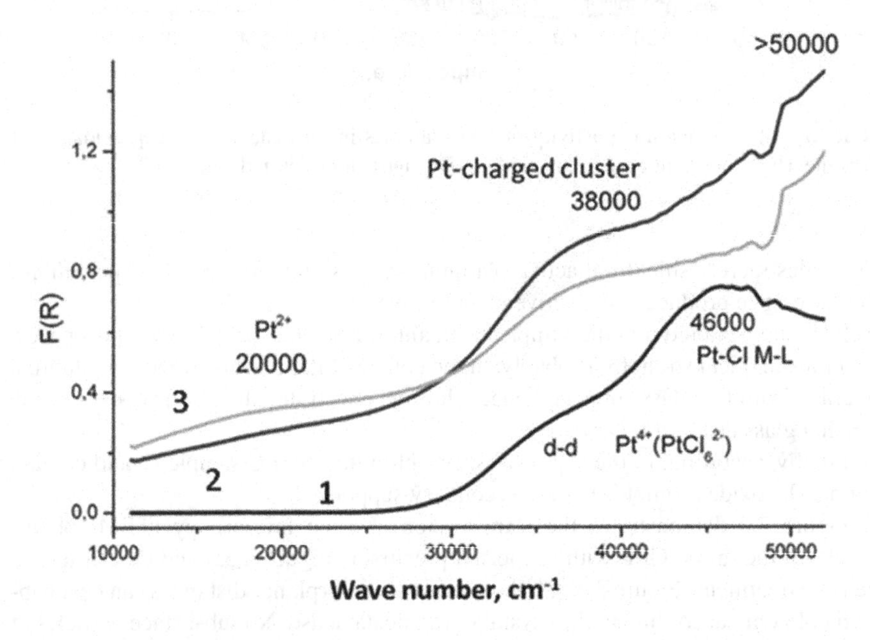


FIGURE 2.4 Electron diffusion reflectance spectrum of the initial glass-fiber fabric with the supported precursor (curve 1) and Pt/GFC after the thermal treatment at 400°C (curve 2) and 700°C (curve 3).

50,000 $cm^{-1}$). The same Pt forms are found in the sample after the high-temperature treatment, but the amount of charged $Pt^{\delta+}$ clusters is lower, and the fraction of ionic Pt and large surface particles is growing. At the same time, these changes are moderate, and it ensures the potentially high stability of the prepared catalyst at high temperatures.

Similar Pt states were observed earlier for the Pt/Zr/GFC of the previous generation.[29] Notably, in the mentioned paper the bands 38,900 $cm^{-1}$ were attributed to the highly dispersed (up to 1 nm) "subsurface" Pt particles. Most probably, this attribution was erroneous, because the formation of any subsurface Pt forms in the synthesized Pt/GFC sample is impossible due to the specifics of the used glass-fiber support with a dense bulk structure without the internal structural voids, which are considered necessary for the formation of the subsurface particles. Besides, it is directly disproved by the electron microscopy data, which have shown a total absence of any platinum subsurface particles in Pt/GFC.

Figure 2.5 shows the data on the comparative activity of different Pt catalysts in the test reaction of deep oxidation of ethylbenzene in air. The comparison involves the commercial conventional Pt catalyst with granular alumina support (~0.56% mass Pt), Pt/Zr/GFC sample prepared by the method described by Balzhinimaev[29] (0.02–0.03% mass Pt at zirconia-silica glass support) and GFC IC-12-S111 (0.07–0.08% mass Pt), synthesized by STS method. The experimental data for the first two samples were taken from the paper of Balzhinimaev.[29]

The data for IC-12-S111 GFC were obtained in the direct flow reactor, using the prismatic GFC cartridge, accurately reproducing the internal geometry of the commercial cartridge (the structure of cartridges and the experimental procedure are described in Chapters 3 and 4 in detail). GFC loading to the cartridge was equal to 6.3 g. The oxidation was performed with the model mixture, containing 100 ppmv of ethylbenzene in the air, it provided a low heat effect of reaction and, therefore, the minimum temperature gradients in the reactor. The flow rate of the reaction mixture was set at the level of 10 liters per minute per 1 g of the catalyst for all the tested catalysts. Adequate measures were also taken to provide the minimization of the diffusion limitations.

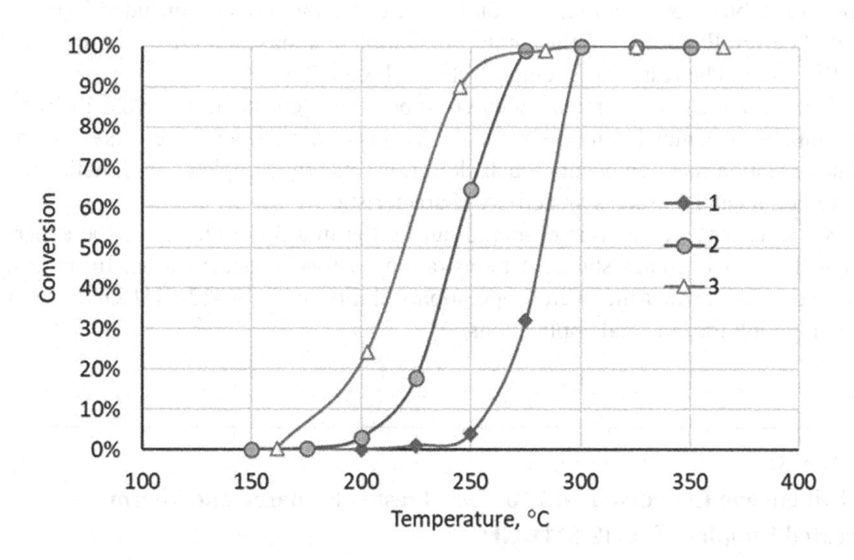


**FIGURE 2.5** Dependence of ethylbenzene conversion on temperature for different catalysts: (1) 0.56% $Pt/Al_2O_3$, (2) 0.02% Pt/Zr-Si GFC, and (3) IC-12-S111 GFC.

*Source:* Reprinted with permission from Lopatin S.A., et al. 2015.[46]

Evidently, the observed activity of both GFCs is much higher than in case of the conventional alumina-based catalyst. Among tested GFCs, some higher activity is demonstrated by the IC-12-S111 catalyst, synthesized by means of the STS method.

The IC-12-S111 catalyst was tested in high-temperature processes of catalytic combustion of gaseous (propane-butane)[91] and solid (dispersed coal) fuels.[99] The operation temperature in these processes was in the range from 650–700°C up to 800–850°C. GFCs have demonstrated a stable operation in these conditions for a few dozen hours, maintaining the effective performance, mechanical strength, and geometric shape.

More accurate quantitative data on the thermal stability of GFC were collected in the experiments, which included short (30 minutes) and long-term (12 hours) incineration of the catalyst samples at 700°C. GFC activity before and after such treatments was measured at 250°C in the direct flow reactor at the test reaction of ethylbenzene oxidation with the use of the model reaction mixture (100 ppmv of ethylbenzene in air). To provide maximum accuracy of the activity measurements, the gas flow rate was set at the value providing ethylbenzene conversion at the fresh GFC in the range 50–70%. The experimental findings are shown in Table 2.1.

It is seen that high-temperature treatment at 700°C produces insignificant change in GFC activity, and the difference in measured conversion is found to be within the measurement error. In summary, the thermal stability of IC-12-S111 Pt/GFC catalyst can be considered quite high.

Lifetime resource testing of this Pt/GFC was performed using the prismatic cartridge in the direct flow reactor with the model CO-propane-air mixture (inlet concentrations: ~140 ppm CO and ~260 ppm propane in air) at residence time of 0.4 seconds, under cyclic alteration of the catalyst heating up to 490°C and its cooling down to ambient temperature. The total resource testing program included 25 cycles with the overall duration of the catalyst operation at a maximum temperature equal to 100 hours. The results are demonstrated in Figure 2.6.

The present CO conversion appeared to be almost constant; it is close to 100%. Presumably, it is mostly limited by the mass transfer limitations. Vice versa, the propane oxidation reaction occurs much slower, practically completely in kinetic area; therefore, it can be used as an activity characteristic.

Some activity decrease is observed during the first 30–40 hours of the experiment. Later it becomes stable, demonstrating no further deactivation. In general, we may consider that the overall operation stability of the IC-12-S111 GFC is sufficiently high for practical applications.

**TABLE 2.1**
**Ethylbenzene Conversion at 250°C for Freshly Prepared and Thermally Treated Samples of IC-12-S111 GFC**

| | Fresh | After 30 minutes incineration at 700°C | After 12 hours incineration at 700°C |
|---|---|---|---|
| Conversion | 58.5% | 57.5% | 57.2% |

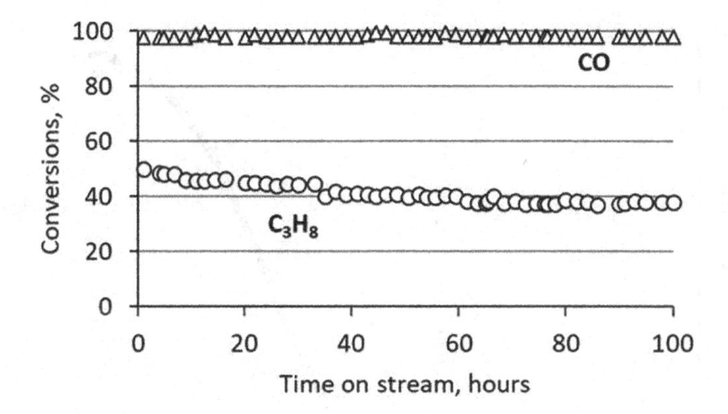


**FIGURE 2.6** Change of CO and propane conversions of IC-12-S111 catalyst at 490°C in course of the resource tests.

*Source:* Reprinted with permission from Lopatin S.A., et al. 2015.[46]

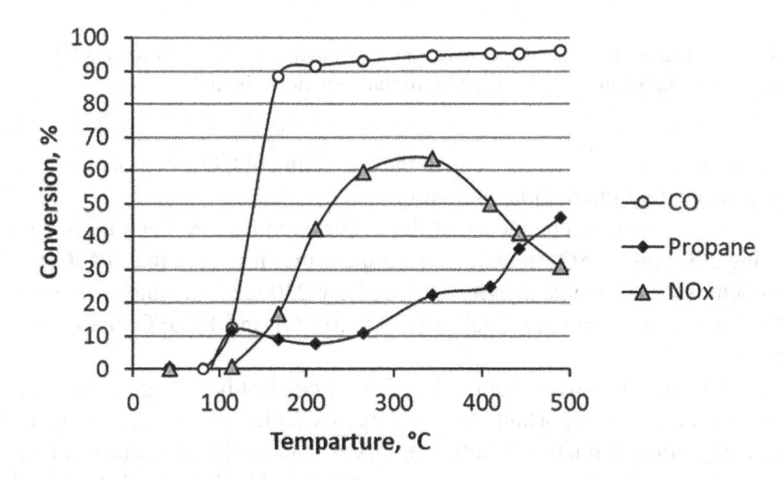


**FIGURE 2.7** Dependence of reactants conversion on temperature in the reaction of catalytic reduction of $NO_x$ by CO and propane in the presence of oxygen. Mass of charged GFC is 48.5 g; initial mixture flow rate is 750 mL/sec.

Figure 2.7 demonstrates the experimental data for the reaction of the catalytic reduction of $NO_x$ by CO and hydrocarbons in the presence of oxygen. The experiments were performed with the initial reaction mixture containing 130 ppm CO, 120 ppm NO, and ~250 ppm propane in air.

According to the experimental results, CO is oxidized at the moderate temperatures—even at 150°C the conversion exceeds 90%. At higher temperatures conversion is increasing more slowly and doesn't achieve 100%, most probably due to the external diffusion limitations. Again, the propane oxidation occurs much slower. $NO_x$ conversion reaches the maximum value (~65%) at ~350°C, decreasing

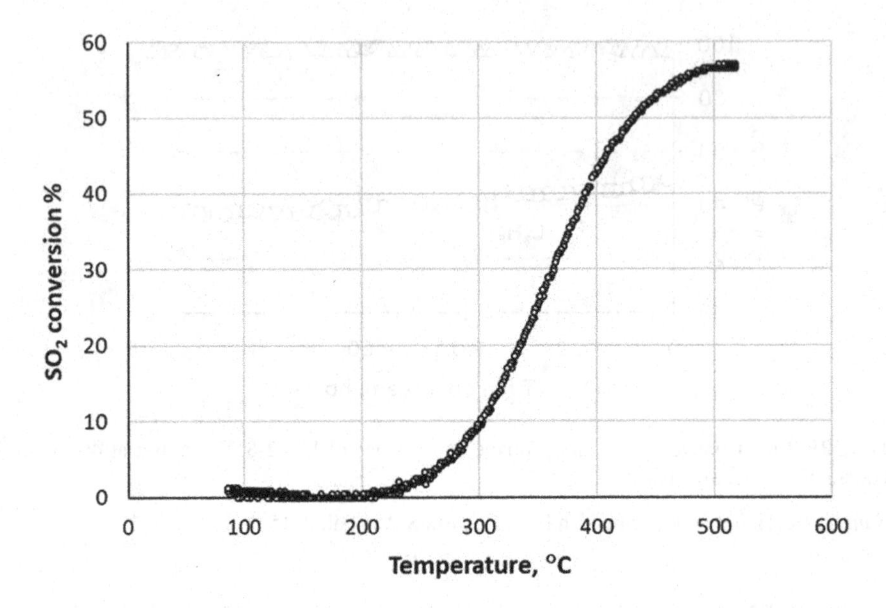


**FIGURE 2.8** Dependence of $SO_2$ conversion on temperature. Mass of loaded GFC is 9 g, reaction mixture flow rate is 22 L/min, $SO_2$ initial content is 200 ppm.

with further temperature increase due to the prevailing of CO and propane oxidation by oxygen over their interaction with $NO_x$.

Figure 2.8 shows the dependence of the sulfur dioxide conversion in the mixture containing 200 ppm of $SO_2$ in the air on temperature. It is seen that the IC-12-S111 catalyst demonstrates visible activity starting from 230–250°C, which is much lower than for traditional vanadium catalysts (360–400°C) and Pt/GFC of the previous generation (~300°C).[49]

In general, we summarize that the IC-12-S111 catalyst has a high catalytic activity in such practically important reactions like oxidation of CO, deep oxidation of organic compounds, reduction of nitrogen oxides, and oxidation of sulfur dioxide. It is also characterized with a high operation stability and high thermal stability, making promising its application in different processes of environmental protection and environmentally friendly combustion of fuels.

## 2.4. COPPER-CHROMITE GFC FOR THE DEEP OXIDATION OF ORGANIC COMPOUNDS

During the last few decades, numerous research efforts have been made to develop new catalysts, with the replacement of expensive noble metals with cheaper active components, such as the transient metal oxides. These oxides are much cheaper, but at the same time they are usually less active; therefore, their higher content (up to few weight percent) must be provided to ensure the overall high catalyst activity. As mentioned earlier, it is necessary to enhance the specific surface area of the glass fibers by covering them with the external layer of the porous secondary support in order to provide the reliable fixing of the active component at the glass surface in the necessary amount,.

Zazhigalov et al.[44] considered the oxide-based GFC for the deep oxidation of organic compounds. Copper chromite, $CuCr_2O_4$, which is characterized by high activity and operation stability in such reactions,[100–104] was chosen as the active component.

The high-silica GFF KT-11-TO (produced by Stekloplastik Co., Zelenograd, Russia), containing ~95% $SiO_2$ and 4% $Al_2O_3$, was selected as the primary GFC support. This fabric has a thermal stability limit as high as 1100°C.

We used $SiO_2$ with the thermal expansion coefficient close to primary glass as the secondary support, thus providing the thermal and mechanical stability of the modified support. The secondary porous silica layer was made by the incipient impregnation of the initial GFF by silica sol (type K-1, produced by NPO IREA, Moscow, Russia). The amount of the supported $SiO_2$ was equal to ~15% of the initial GFF weight (for a dry basis). After impregnation the samples were dried at ambient temperature and thermally treated in the air medium. This procedure has allowed for an increase in the unit surface area of the support from 1 $m^2/g$ for the initial GFF up to 20–30 $m^2/g$.

Then the modified support was impregnated by the water solution of the copper chromite precursors. One of the precursors was copper dichromate; the second one was a mixture of copper nitrate as a copper source and glycine as a combustible additive.

After the impregnation and drying the catalyst samples were incinerated in air medium according to the STS procedure at fast temperature rise from the ambient up to 500°C.

The proposed method provides a way to deposit the copper chromite on the modified GFF in a significant amount, for example, up to 4.5% Cu and 6.0% Cr. Theoretically, the method with the application of copper dichromate as a precursor may provide even higher loading of the copper chromite in the GFC product, but it is hardly reasonable, as it may result in a complete filling of the catalyst pores by the active component and lower the catalyst-active surface area, which, in turn, would decrease the active component availability for the reactants, thus lowering GFC activity.

According to the elemental analysis data (Table 2.2), the dichromate-based sample (No. 1) has the mass relation of copper and chromium content, which equals 0.6, strongly corresponding to the stoichiometric value for $CuCr_2O_4$. This value for the nitrate-glycine sample (No. 2) is much higher (~1.2). It means that this catalyst contains the excess copper oxide in addition to the copper chromite.

The conventional commercial-granulated copper-chromite catalyst ICT-12–8 ($CuCr_2O_4/Al_2O_3$; produced by Katalizator Co., Novosibirsk, Russia) was used in the experiments as a reference sample for the comparison. The properties of all the tested catalyst samples are summarized in the Table 2.2.

X-ray diffraction (XRD) patterns of GFCs have shown the presence of a $CuCr_2O_4$ phase (structural type of spinel, space group *I4₁/amd*), with some presence of $\alpha$-$Cr_2O_3$ (structural type of $\alpha$-$Al_2O_3$, space group *R3c*) and CuO (space group *C2/c*) phases.

The XRD pattern of the traditional aluminum-copper chromium catalyst contains only reflexes of $CuCr_2O_4$ and $\alpha$-$Al_2O_3$. The unit cell parameters of the active

**TABLE 2.2**
**Properties of the Synthesized and Reference Catalysts for VOC Oxidation**

| Sample no. | Type | Precursor | Content of active metal, % mass | | Unit surface area, $m^2/g$ | Pore volume, $cm^3/g$ | Loading in the experimental reactor, g |
|---|---|---|---|---|---|---|---|
| | | | Cu | Cr | | | |
| 1 | Cu-Cr/GFC | $CuCr_2O_7$ | 2.06 | 3.42 | 11 | 0.016 | 10.3 |
| 2 | Cu-Cr/GFC | $Cu(NO_3)_2$, $Cr(NO_3)_3$, glycine | 2.95 | 2.5 | 20 | 0.04 | 10.2 |
| 3 | Cu-Cr/$Al_2O_3$ | — | 5.0 | 7.6 | 112 | 0.25 | 43.0 |

*Source:* Reprinted with permission from Zazhigalov S., et al. 2017.[44]

**TABLE 2.3**
**The XRD Data**

| Sample no. | Active component | Parameters of unit cell | | | β | V, Å³ | sp. gr. |
|---|---|---|---|---|---|---|---|
| | | a | b | c | | | |
| 1 | | 6.024 | – | 7.828 | 90 | 281 | *I4₁/amd* |
| 2 | $Cu(CrO_2)_2$ | 6.018 | – | 7.827 | 90 | 286 | *4₁/amd* |
| 3 | | 6.022 | – | 7.834 | 90 | 284 | *I4₁/amd* |
| ICSD data[107] | | 6.034 | – | 7.788 | 90 | 283 | *I4₁/amd* |

*Source:* Reprinted with permission from Zazhigalov S., et al. 2017.[44]

component are in agreement with the parameters of ICSD database[105] within the margins of XRF error (Table 2.3).

The average crystallite sizes calculated from the Scherrer equation[106] were equal to 17 nm for the GFC sample No. 1, 12 nm for GFC sample No. 2, and 29 nm for the conventional $Al_2O_3$-based catalyst (sample No. 3).

The UV-Vis DR spectra for the tested catalysts are given in the Figure 2.9. Copper chromite, modified by cations $Me^{3+}$ (e.g., $Fe^{3+}$), has only two absorption bands, at 13,600 and 16,700 $cm^{-1}$, in the visible range of UV-Vis DR spectrum.[108] These bands are caused by the expression of d–d-transitions of $Cu^{2+}$ and $Cr^{3+}$ cations stabilized in the octahedral oxygen coordination.

According to the UV-Vis DR spectrum, the traditional alumina-based copper chromite catalyst (Figure 2.9, curve 1) contains a $CuCr_2O_4$ phase with a structure of the partially inverted spinel, with $Cu^{2+}$ and $Cr^{3+}$ cations stabilized in the octahedral oxygen coordination.

Intensive absorption bands of 27,000 $cm^{-1}$ and 37,800 $cm^{-1}$ are most probably caused by the expression of the charge transfer bands (CTB) of the ligand-metal $Cr^{3+}$

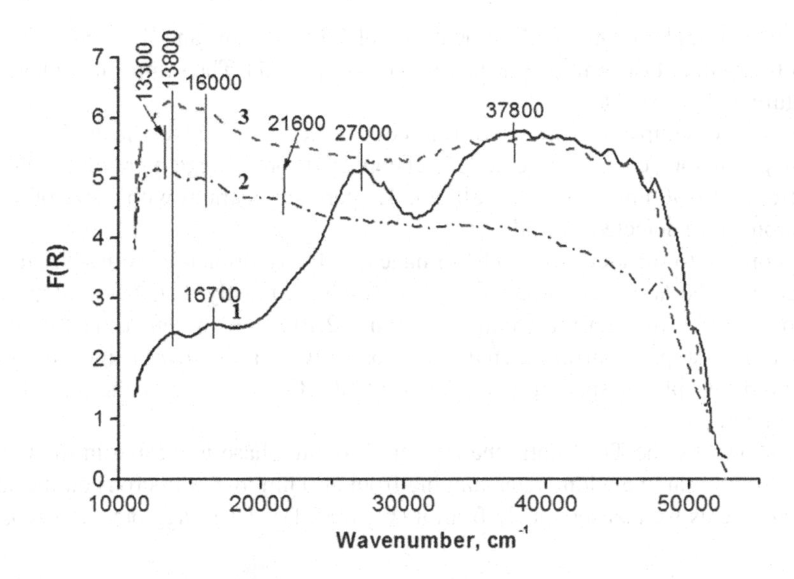


**FIGURE 2.9** UV-Vis DR spectrum of the copper-chromium catalysts: (1) traditional aluminum-copper chromium catalyst, (2) sample No. 1, (3) sample No. 2.

*Source:* Reprinted with permission from Zazhigalov S., et al. 2017.[44]

cations in the octahedral oxygen coordination stabilized in the structure of $\alpha$-$Al_2O_3$, which can dissolve up to 8 atomic % Cr.[109]

The UV-Vis DR spectrum of the synthesized GFCs (Figure 2.9, curves 2, 3) contain an absorption band of 16,000 $cm^{-1}$ caused by d–d-transition of $Cr^{3+}$ cations into tetragonal-distorted octahedral coordination.[109,110] The second absorption band is observed at 13,800 $cm^{-1}$ for the sample No. 1 (Figure 2.9, curve 2) and at 13,300 $cm^{-1}$ for sample No. 2 (Figure 2.9, curve 3). These bands are highly likely caused by the expression of d–d-transition $Cu^{2+}$ cations into tetragonal-distorted octahedral coordination. Slightly different values of the absorption band energy for $Cu^{2+}$ cations, stabilized in octahedral positions of $CuCr_2O_4$ spinel, indicate a decrease of the tetragonal distortion degree for the oxygen octahedral around $Cu^{2+}$ cations at the transition from sample No. 1 to sample No. 2. It means that the value of the tetragonal distortion of the oxygen octahedron around $Cu^{2+}$ cations strongly depends on the methods of $Cr^{3+}$ and $Cu^{2+}$ cation deposition at the modified glass-fiber surface and on their precursors' nature. It is possible to make qualitative evaluation of the fraction of spinel $CuCr_2O_4$ inversion from the intensity of the absorption band at 13,300–13,800 $cm^{-1}$. The number of $Cu^{2+}$ cations stabilized in the octahedral sites of the $CuCr_2O_4$ spinel structure increases from sample No. 1 to sample No. 2. Therefore, the method of $Cr^{3+}$ and $Cu^{2+}$ cation deposition produces a significant effect on the interaction between these metals during formation of the active component. Evidently, the undesirable formation of the $\alpha$-$Cr_2O_3$ phase on the support surface is caused by the excess of chromium precursor at the catalyst synthesis stage.

According to the scanning-electron microscopy (SEM) data, the active component $CuCr_2O_4$ is mainly located in the traditional alumina-based catalyst on the alumina surface as agglomerates with a size in the range of 1–4 microns. Smaller oval-shaped agglomerates with dimensions of 250–500 nm and 400–600 nm are also observed, but their amount is not high (less than 5%). The sample has a layered structure without voids.

Both GFC samples, synthesized from copper dichromate and copper/chromium nitrate precursors, contain the oblong oval copper chromite agglomerates ranging in size from 60 × 90 nm up to 180 × 210 nm. Larger agglomerates with a size of up to 2 microns were detected as well.

According to the analysis of TEM images of the traditional alumina-based Cu/Cr catalyst, the active component of this catalyst is mostly distributed along the support surface in a uniform manner (Figure 2.10), though sometimes the concentration of copper chromite changes across a wide range (from light to dark). Observed interplanar spacing is typical for $CuCr_2O_4$. The size of the particles is 100 nm and more.

As shown by the TEM data, the copper chromite phase is located in the GFCs in the form of large agglomerates ranging from 500 nm to 2–3 microns on the support surface as well as separately from it (Figure 2.11). These agglomerates mostly

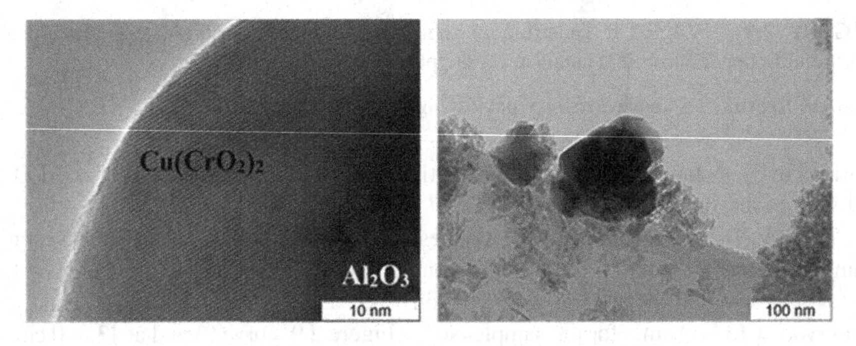


**FIGURE 2.10** Electron microscopic image of the traditional alumina-based Cu/Cr catalyst.
*Source:* Reprinted with permission from Zazhigalov S., et al. 2017.[44]

**FIGURE 2.11** Electron microscopic images of Cu/Cr GFC.

*Source:* Reprinted with permission from Zazhigalov S., et al. 2017.[44]

consist of particles with the size from 10–15 nm to 20–25 nm. In both GFC samples the active component particles are also detected in the subsurface layers in the bulk silica matrix at a depth up to 3–5 nm.

The reaction of the deep oxidation of toluene in the air was used to compare the catalytic activity of the prepared catalysts. The inlet toluene concentration in the inlet air was rather low (~120 ppmv) to minimize the adiabatic heat rise of the reaction and thus to provide efficient temperature control inside the lab reactor. This reaction mixture was prepared by the controlled mixing of a small flow of helium, saturated with toluene vapors, with the main flow of air. Toluene concentration at the reactor inlet and outlet were measured by the means of the gas chromatograph Tsvet-500M with flame-ionization detector.

The total flow rate of 27 std.l/min was maintained during the experiments. The reaction temperature changed within the range from 20–500°C. All experiments were performed at ambient pressure.

The flow reactor, used in experiments, had the two-zone heating external chamber and internal mixing zone, which contains the steel spheres, providing for the uniform distribution of the reaction fluid across the reactor sequence. All catalyst samples were charged into the reactor in equal volume (64 $cm^3$). The GFCs (Nos. 1 and 2) were charged in the form of a cubic structured cartridge. The traditional catalyst (No. 3) was loaded in the form of commercial ring-shaped granules with a size 10 × 10 × 2.5 mm, also placed inside the permeable cubic steel wire mesh cartridge. It allowed to bring the test conditions to the conditions of the commercial catalytic process as close as possible to make relatively accurate reproduction of the mass transfer limitations.

The dependence of the toluene conversion upon the reaction temperature is shown in Figure 2.12. It is seen that both GFC samples demonstrate significantly higher activity than that for alumina-based catalyst. Among the GFCs, the sample No. 2 seems to be slightly more active, most probably due to the higher total copper content.

The apparent reaction rate constant was used to estimate the catalytic activity, calculated on the basis of the assumption of the plug flow regime in the experimental reactor and the first-order reaction rate equation with respect to the concentration of toluene. As seen from Table 2.2, the apparent densities of the catalyst beds in case of the GFCs and the traditional catalyst pellets were significantly different; therefore, to ensure the objective comparison of GFCs and a traditional catalyst with a very different packing density, the specific rate constants were calculated per unit volume of the catalyst bed,

$$k_V = -\frac{Qln(1-x)}{V} \qquad \text{sec}^{-1}$$

and per unit mass of the active component,

$$k_m = -\frac{Qln(1-x)}{MC_{Cu}} \qquad \text{std.l/g/sec}$$

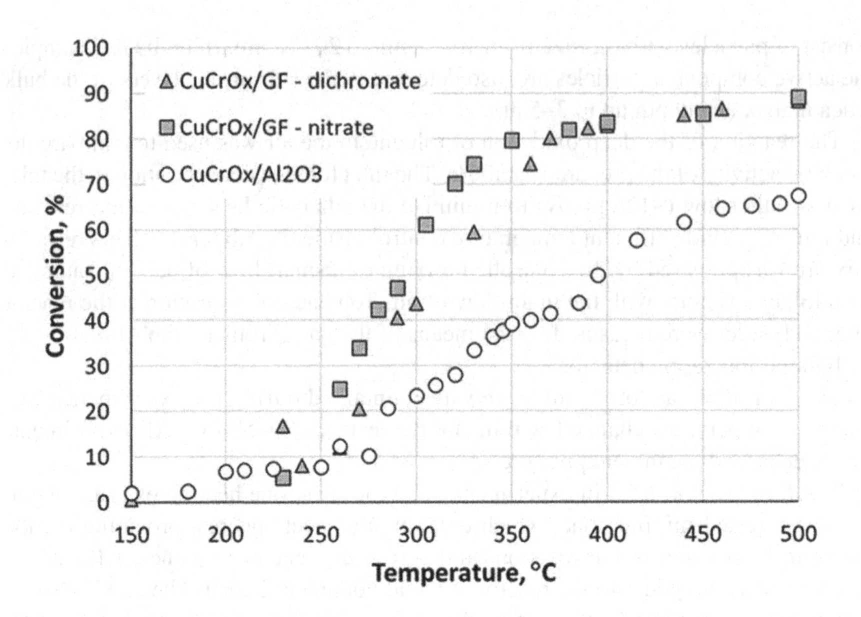


**FIGURE 2.12** Dependence of toluene conversion on temperature. Mixture: 120 ppm toluene in the air, ambient pressure, GHSV = 25,000 $h^{-1}$.

*Source:* Reprinted with permission from Zazhigalov S., et al. 2017.[44]

where $Q$ is the volume flow rate of the reaction mixture (std.l/sec), $V$ is the catalyst bed volume (l), $x$ is the conversion, $M$ is the mass loading of the catalyst in the reactor (g), and $C_{Cu}$ is the mass fraction of copper in the catalyst (mass fraction).

The apparent rate constants calculated per unit volume of the catalytic cartridge (Figure 2.13a) follow the conversion behavior observed in Figure 2.12. The substantial changes are seen for the constants calculated per unit mass of the catalyst (Figure 2.13b). Both GFCs show practically equal unit activity, and the specific activity of the copper chromite in these samples significantly exceeds that for alumina-based sample, because of much lower GFC mass in the cartridge. This difference starts from 3 times in the area of low temperatures and reaches 20–30 times at higher temperatures.

We may assume that the $CuCr_2O_4$ in the GFCs has higher intrinsic activity than that for alumina-based catalyst; it is most probably connected with the smaller particle size of the copper chromite in GFCs. It leads to the higher observed activity at low temperatures with the reaction occurring in a kinetic region. At higher temperatures the difference of the activity levels increases, obviously due to a higher efficiency of both the internal and the external mass transfer in GFC cartridges.[111]

As mentioned earlier, some fraction of the active component is located under the surface of the secondary support, inside the silica layer volume. To some extent this is similar to subsurface active component species, reported earlier in respect to Pt/GFC.[29] As the existence of such subsurface active centers was

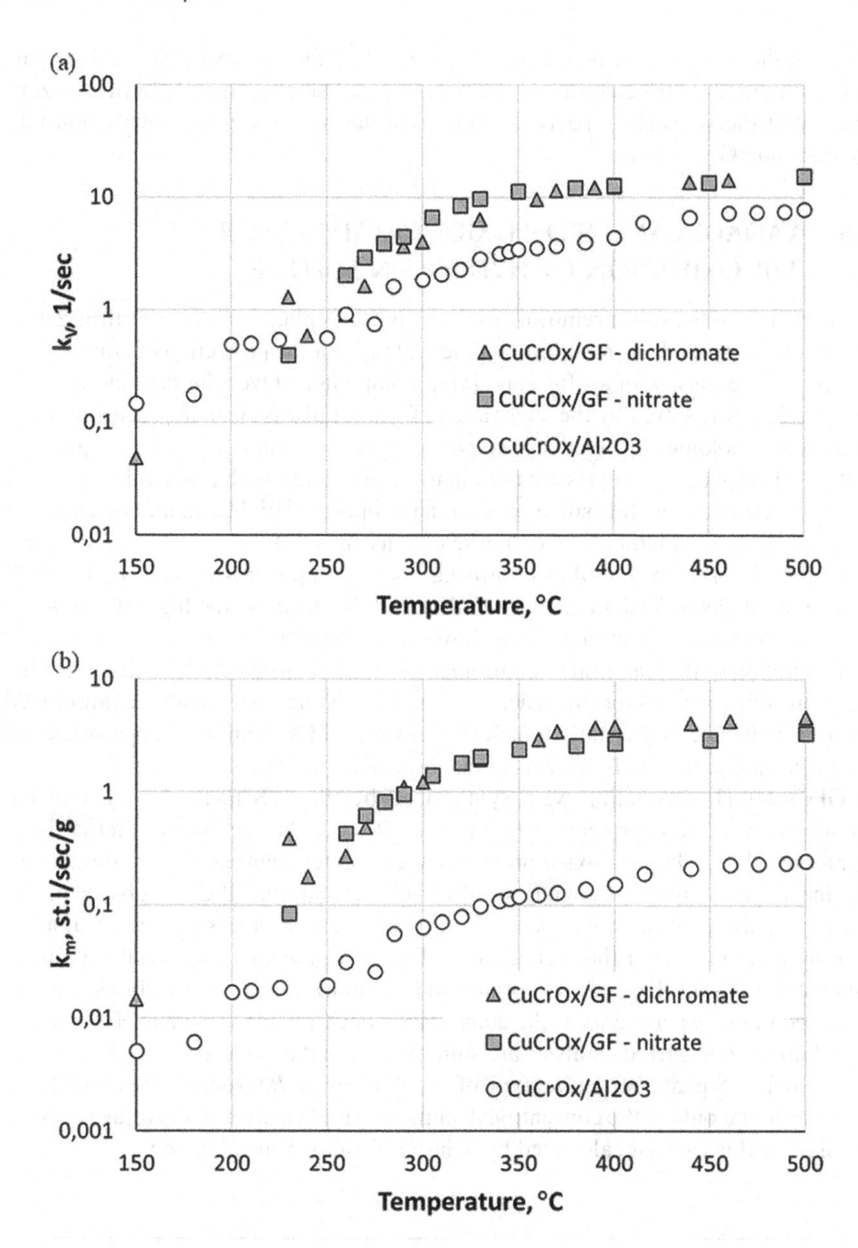


**FIGURE 2.13** Temperature dependence of the specific apparent rate constants calculated per unit catalyst bed volume (a) and per unit mass of active component in the bed (b).

*Source:* Reprinted with permission from Zazhigalov S., et al. 2017.[44]

doubtful, it was interesting to evaluate the possible contribution of such subsurface copper chromite into overall GFC activity. We have synthesized a catalyst sample in which the active component particles were completely removed by washing them with nitric acid from the external surface of fibers. Subsurface particles remained unaffected during the treatment. The experiments indicated zero activity of the subsurface particles, thus postulating their zero contribution into the common GFC activity.

## 2.5. VANADIA AND IRON OXIDE CATALYSTS FOR THE OXIDATION OF HYDROGEN SULFIDE

Improvement of the sulfur removal efficiency in Claus plants, used in desulfurization of natural gas and oil, is an important practical task for the protection of atmospheric air from the emission of sulfur-containing compounds. Over the last few decades this problem was solved by the application of Claus tail gas cleanup processes; some of these technologies involved a selective catalytic oxidation of $H_2S$ by oxygen into sulfur.[112] $FeO_x$-based catalysts are used in these processes, with iron oxide supported on a granular or monolith silica or alumina supports.[113,114] The main advantage of these catalysts is a high $H_2S$ oxidation selectivity into sulfur.

Except for iron oxide, other promising active components can be used in $H_2S$ oxidation catalysts. As demonstrated earlier,[115,116] $V_2O_5$ shows the highest activity in $H_2S$ oxidation reaction among all the known metal oxides.

In addition to the traditional requirements for the catalyst activity and selectivity, the promising catalysts for this reaction should be characterized with the minimized limiting influence of the mass transfer processes. GFCs seem most appropriate for the given application both in terms of catalytic and mass transfer properties.

GFCs for $H_2S$ oxidation were synthesized by the STS method, including the modification of the fiber support by the deposition of the secondary external layer of porous silica. The iron oxide precursor was a water solution of iron nitrate with glycine as fuel additive.[57] A synthesis of vanadia-containing GFCs was based on the use of vanadyl oxalate water solution[54–56]; oxalate anion in this precursor simultaneously played a role of the fuel additive. The conventional commercial iron oxide catalyst ICT-27–42 ($Fe_2O_3/Al_2O_3$, produced by Katalizator Co., Novosibirsk, Russia) was used in the experiments as the comparative $FeO_x$ reference sample. The properties of all the considered catalysts are summarized in the Table 2.4.

The $\alpha$-$Fe_2O_3$ phase (structural type of $\alpha$-$Al_2O_3$, sp. gr. *R3c*) according to the XRD data was found only in the conventional alumina-based catalyst. The average crystallite size for this catalyst calculated by Scherrer formula[107] is 22–24 nm.

**TABLE 2.4**
**Properties of the Synthesized and Reference Catalysts for $H_2S$ Oxidation**

| Sample | Type | Active component (mass content) | Unit surface area, $m^2/g$ |
|---|---|---|---|
| No. 1 | $Fe_2O_3$-$SiO_2$/GFC | $Fe_2O_3$ (1.9% Fe) | 30 |
| No. 2 | $V_2O_5$-$SiO_2$/GFC | $V_2O_5$ (5.4% V) | 9.5 |
| No. 3 | $Fe_2O_3/Al_2O_3$ | $Fe_2O_3$ (~10% Fe) | 113 |

This phase was not detected in Fe-based GFC, possibly due to the transformation of $\alpha$-$Fe_2O_3$ into the amorphous form or into the species with a very small particle size (less than 2 nm).

In the visible region of UV-Vis DR spectrum of the conventional iron oxide catalyst (Figure 2.14, curve 1) it is possible to see the absorption band with a low intensity and with a maximum peak at 11,600 $cm^{-1}$, accompanied with a shoulder at 15,000 $cm^{-1}$ caused by the appearance of d–d-transitions of $Fe^{3+}$ cations in the oxygen octahedral coordination ($Fe^{3+}{}_{Oh}$), stabilized in the oxide matrix of $\alpha$-$Fe_2O_3$.[117]

The high-intensity absorption, observed at the ultraviolet spectral area, is caused by the appearance of the bands, corresponding to charge transfer of the ligand-metal cations $Fe^{3+}{}_{Oh}$, which often conceal *d–d*-transitions of cations $Fe^{3+}{}_{Oh}$, usually having lower intensity.

It has to be mentioned that $FeO_x$/GFC spectrum is different from that of conventional $FeO_x$ catalyst containing the well-crystallized $\alpha$-$Fe_2O_3$ phase, as shown by the XRD data. The spectrum of GFC sample (Figure 2.14, curve 2) shows a strong absorption in the area above 15,000 $cm^{-1}$, which is a characteristic of the charge transfer bands of the ligand-metal cations in the $Fe^{3+}{}_{Oh}$ ligands' oxygen environment.[117]

We may assume that the GFC sample contains the $\alpha$-$Fe_2O_3$ particles with the size less than 5 nm on the support surface. Cations of $Fe^{3+}{}_{Oh}$ are located on the catalyst surface in an isolated state and presumably stabilized in the surface layers of the $SiO_2$ secondary support.

According to SEM data, the active component agglomerates in the conventional $FeO_x/Al_2O_3$ catalyst have mainly an oval shape with a size from 250–300 nm to 2–4 microns (Figure 2.15a).

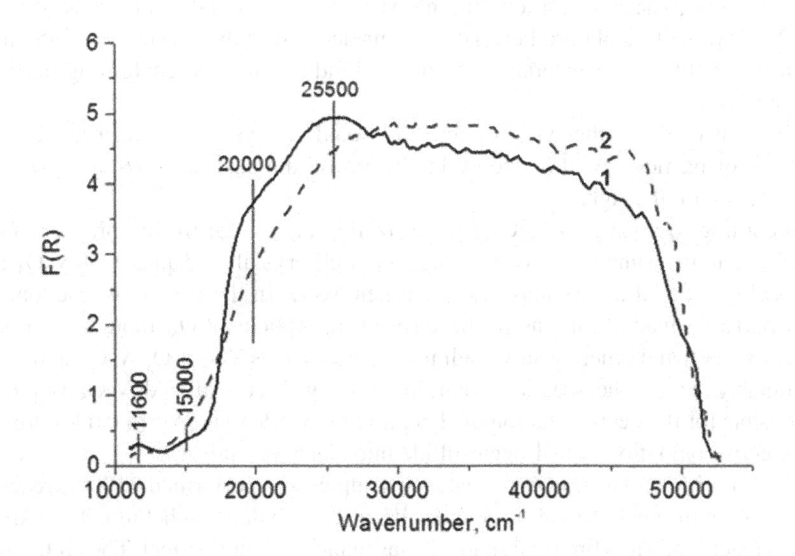


**FIGURE 2.14** UV-Vis DR spectrum of the iron samples: (1) conventional $FeO_x/Al_2O_3$ catalyst and (2) $FeO_x$/GFC.

*Source:* Reprinted with permission from Mikenin P., et al. 2016.[57]

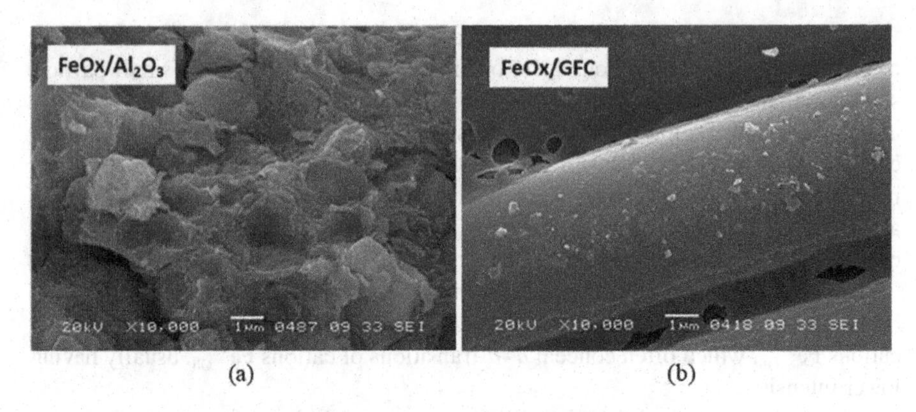


**FIGURE 2.15** The electronic image of the conventional $FeO_x/Al_2O_3$ catalyst (a) and $FeO_x$/GFC (b) in secondary electrons.

*Source:* Reprinted with permission from Mikenin P., et al. 2016.[57]

$FeO_x$/GFC also contains small and medium oval-shaped agglomerates ranging in size from 50 to 400 nm (Figure 2.15b). Complex shapes with well-defined borders are also observed.

As shown by TEM images of the $FeO_x$/GFC sample (Figure 2.16, on the left), silica is localized in the form of large porous particles at the surface of the glass fibers as well as separately from them. Iron oxide is mostly present in the form of flattened $Fe_2O_3$ islets with a longitudinal size of 1–2 nm and a thickness of 0.5–1 nm (2–3 layers of octahedra $FeO_6$) on the surface and in the subsurface (4–5 nm in depth) layers of silica secondary support. Individual large Fe-containing particles were not found.

The iron oxide in the $Al_2O_3$-based traditional catalyst is present preferably in the form of particles with a size of 15–30 nm on the surface of $\alpha$-$Al_2O_3$ support (Figure 2.16, on the right).

According to a vanadia GFC characterization data,[58] the freshly prepared GFC sample contains vanadia in two forms: the well-crystallized phase of $V_2O_5$ and the weakly ordered amorphous vanadium pentoxide. In the course of reaction, the main part of vanadia remains in the form of amorphous $V_2O_5$, though some part reduces to $VO_2$ and other lower vanadium oxides, such as VO, $V_2O_3$, $V_3O_7$, and $V_4O_9$. Presumably, under the reaction conditions, the well-crystallized $V_2O_5$ is mostly responsible for the deep oxidation of $H_2S$ into $SO_2$, while the reduced oxides provide the selective oxidation of hydrogen sulfide into elemental sulfur.

The experimental testing of the catalyst samples was performed with the reaction mixture containing 1.0% vol. $H_2S$ and 1.1% vol. $O_2$, balance—helium. The experimental reactor had a cylindrical shape; it was made of stainless steel. The GFCs were charged into the reactor in a form of the spiral cartridge, where the catalyst fabric was rolled in a combination with the corrugated structuring steel foil.

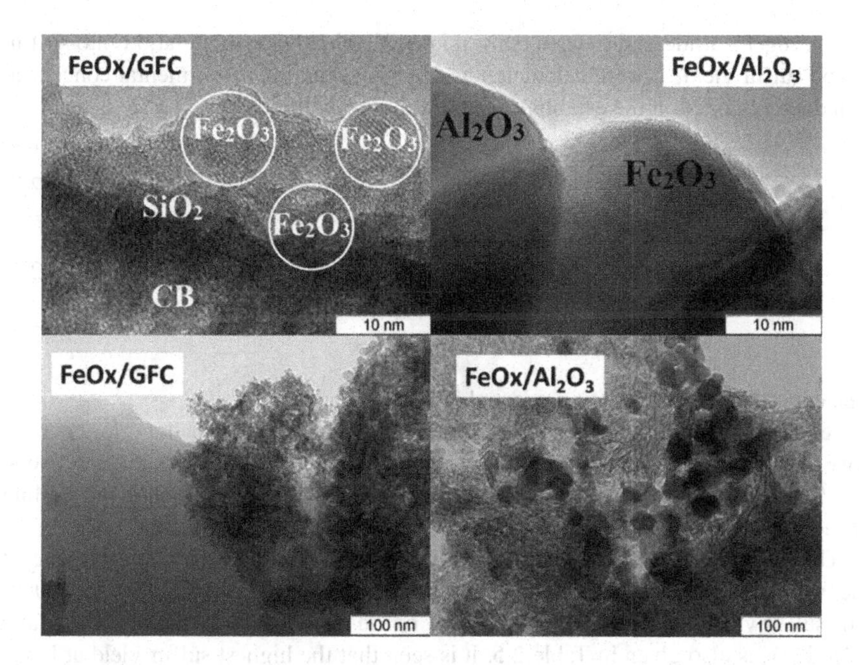


**FIGURE 2.16** TEM images of the FeOx/GFC (left) and traditional FeOx/$Al_2O_3$ catalyst (on the right).

*Source:* Reprinted with permission from Mikenin P., et al. 2016.[57]

The traditional granular $FeO_x$ catalyst, initially manufactured in the form of cylinders 10 mm in diameter and 5–7 mm in height, was milled into the pieces less than 1 mm prior to charging into the reactor to minimize the diffusion limitations.

The flow rate of the inlet reaction mixture for all catalysts was set equal to 1.5 mL/sec per 1 g of the catalyst. The temperature in the reactor varied in the range from 175°C to 250°C with the step of 25°C. The experiment at each temperature level was performed until the confirmed steady state. It was determined from the achievement of the constant reagent conversion for at least 2 hours and the attainment of zero (within the concentration measurement error) mass balance deviation (δ), calculated as

$$\delta = \Delta C_{O_2} + C_{SO_2}^{out} - \frac{1}{2}\Delta C_{H_2S} \tag{2.1}$$

where $\Delta C_{O_2}$ and $\Delta C_{H_2S}$ are the differences between the inlet and outlet oxygen and hydrogen sulfide concentrations and $C_{SO_2}^{out}$ is the outlet $SO_2$ concentration (vol.%).

All the experiments were performed at an ambient pressure. The analysis of the inlet and outlet reaction mixtures was carried out by the means of the gas chromatograph Tsvet-500M.

Hydrogen sulfide and oxygen conversions, $X$, selectivity of oxidation into sulfur, $S$, and sulfur yield, $Y$, were calculated as follows, using the experimental concentration data, obtained in the established steady-state conditions:

$$X_i = \frac{\Delta C_i}{C_i^{in}} \tag{2.2}$$

$$S = 1 - \frac{C_{SO_2}^{out}}{\Delta C_{H_2S}} \tag{2.3}$$

$$Y = X_{H_2S} \times S \tag{2.4}$$

The experimental results are presented in Table 2.5.

As it is seen from the experimental data, $FeO_x$ (No. 1) and vanadia (No. 2) GFCs have almost equal activity above 200°C, while the vanadia sample is more active at 175°C. The conventional $FeO_x$ catalyst (No. 3) is much less active, though the loading of the iron oxide in it allows few times higher activity.

The observed selectivity is rather high (more than 90%) for all the catalysts, except the $V_2O_5$ sample, demonstrating the decrease of selectivity to 70–75% at 250°C. The sulfur yield, characterizing the quality of the compromise between activity and selectivity, is also given in Table 2.5. It is seen that the highest sulfur yield at lower

**TABLE 2.5**
**Dependence of $H_2S$ Conversion, Selectivity of Oxidation into Sulfur, and Sulfur Yield upon Temperature for Different Catalysts**

| Temperature, °C | $H_2S$ conversion | $O_2$ conversion | Selectivity into sulfur | Sulfur yield |
|---|---|---|---|---|
| **No. 1 $Fe_2O_3$-$SiO_2$/GFC** | | | | |
| 175 | 32.7% | 15.6% | 98.3% | 32.2% |
| 200 | 53.8% | 26.1% | 96.2% | 51.8% |
| 225 | 79.0% | 44.4% | 92.6% | 73.2% |
| 250 | 95.4% | 61.9% | 90.0% | 85.9% |
| **No. 2 $V_2O_5$-$SiO_2$/GFC** | | | | |
| 175 | 52.8% | 15.5% | 99.9% | 52.7% |
| 200 | 58.5% | 20.4% | 97.8% | 57.2% |
| 225 | 80.5% | 28.9% | 94.8% | 76.3% |
| 250 | 96.9% | 81.8% | 73.4% | 71.1% |
| **No. 3 $Fe_2O_3$/$Al_2O_3$** | | | | |
| 180 | 16.0% | 13.4% | 99.9% | 16.0% |
| 210 | 29.0% | 25.1% | 98.0% | 28.4% |
| 240 | 51.0% | 46.7% | 95.1% | 48.5% |
| 250 | 68.9% | 66.2% | 92.4% | 63.6% |

temperatures is ensured by vanadia GFC No. 2, with the better performance of iron-oxide GFC No. 1 at higher temperatures as a result of the higher selectivity. The traditional alumina-based sample No. 3 shows lower sulfur yield due to a lower activity despite its good selectivity.

The values of the specific rates of elemental sulfur formation at different catalysts calculated per unit mass of catalyst and active component are presented in Table 2.6. These values were measured in the experiments with mixture containing ~1.0% vol. $H_2S$, temperature 225°C, reaction mixture flow rate of 90 mL/min, mass of each catalyst ~1 g.

According to the presented data, the iron oxide (No. 1) and vanadia (No. 2) GFCs have almost equal specific activity per 1 g of the catalyst, but the Fe-based GFC is more active by 2.7 times in respect to 1 g of the active metal.

Besides, the unit activity of the No. 1 GFC is one order of magnitude higher than that of the traditional alumina-based catalyst (No. 3). Presumably, this serious difference in the specific activity of $FeO_x$ catalysts is caused by the difference in the state and nature of the active iron oxide.

It may be due to a much higher dispersion (1–2 nm) of the $FeO_x$ particles in GFC in comparison with the traditional $FeO_x/Al_2O_3$ catalyst, with typical size of the active species in the range of 15–30 nm. The GFC contains iron oxide in the form of the isolated $Fe^{3+}{}_{Oh}$ cations, stabilized in the surface layers of $SiO_2$ secondary support. This form differs from a well-crystallized $\alpha$-$Fe_2O_3$ phase, which is typical for the traditional $Al_2O_3$-based catalyst.

Summarizing, vanadia– and iron oxide–based GFCs both seem appropriate for practical application in the industrial processes of $H_2S$ selective oxidation. The vanadia-based catalyst is more promising for application at lower temperatures (below 200°C) due to its higher activity. In turn, the more selective $FeO_x$-based GFC can be more suitable to be used at higher temperatures. These catalysts may be used either separately or in the combination, making the most of their advantages; for example, vanadia catalyst can be applied in the inlet part of the adiabatic $H_2S$ oxidation reactor, while the FeOx catalyst can be placed in its outlet part, where the temperature is higher.

**TABLE 2.6**

**Comparison of the Sulfur Formation Rates in Different Catalysts**

| Sample | Type | Specific rate of elemental sulfur formation at 225°C, mg/sec | |
|---|---|---|---|
| | | per 1 g of catalyst | per 1 g of active metal |
| No. 1 | $Fe_2O_3$-$SiO_2$/GFC | 0.016 | 0.826 |
| No. 2 | $V_2O_5$-$SiO_2$/GFC | 0.016 | 0.303 |
| No. 3 | $Fe_2O_3/Al_2O_3$ | 0.008 | 0.081 |

*Source:* Reprinted with permission from Mikenin P., et al. 2016.[57]

Notably, the GFC activity was investigated in the given study with the use of the structured cartridge, reproducing the geometry of the commercial-scale catalyst bed, and therefore the mass transfer conditions in the real reactor. This opens the way to a direct measurement of the apparent reaction rates and thus to the significant simplification of the scale-up procedure. The scale transfer for the conventional catalyst should include the transition from the small catalyst particles (less than 1 mm) used in the experiments to the commercial-size pellets with the increase of both internal and external mass transfer limitations resulting in the decrease of the apparent reaction rates. Finally, under the conditions of industrial processes, we may expect even more significant superiority of the $FeO_x$/GFC over the conventional catalyst than was observed during the experiments.

## 2.6. MULTILAYER COMPOSITE MATERIAL WITH THE TERNARY LAYER OF NANOFIBROUS CARBON

Nowadays, materials containing nanoscale carbon structures (e.g., nanofibers, nanotubes, graphene and others) attract considerable interest from researchers and practitioners in many areas. Such materials can be applied in the production of:

- Advanced supercapacitors[118,119]
- Devices for the thermal and photovoltaic conversion of solar energy[120–123]
- Polymers and composite materials with a high thermal and electric conductivity, protective shields for the absorption of electromagnetic waves[124–126]
- Catalyst and catalytic supports[127–129]

and many other brand new and emerging materials.

Carbon nanostructures can be produced, in particular, using the pyrolysis of hydrocarbon feedstock at the catalysts containing the metals that are active in the pyrolysis reactions, such as nickel.[130,131]

Deposition of the carbon film directly on the surface of the glass fibers is complicated due to a low maximum amount of the active metal and its weak binding with the glass surface, which leads to a low mechanical strength of the manufactured material. It is possible to overcome this problem by applying the modification of the glass-fiber support with the additional porous layer of the secondary support, described earlier, while providing a way to deposit larger quantity of the active metal with a stronger surface bonding.

The glass-fiber catalyst with nickel as an active component and modified with porous $Al_2O_3$ as secondary support was used for the catalytic deposition of the carbon nanofibers (CNF).[94] This produced a multilayered material with an external layer of CNF and the specific surface area of ~50 $m^2/g$. This result looks promising, but the application of alumina as the secondary support presents certain difficulties: the $Al_2O_3$-modified GFCs are characterized by lower flexibility; moreover, after the thermal processing of such modified system, our experience has shown that they become quite fragile and lose their mechanical strength significantly. This can be explained by the penetration of alumina into the glass matrix at high temperatures causing destructive changes in the glass structure.

The synthesis of the novel material with the formation of CNF was based on the glass fiber support, pre-modified with the additional layers of silica and nickel oxide.[132–134]

Modification of the initial GFF was made in accordance with the procedure described. The support was initially modified with the additional secondary layer of $SiO_2$ and impregnated with a water solution of Ni acetate, where acetate anion played the role of a fuel additive. Impregnation was followed by drying and incineration at 500°C in air medium.

The produced catalyst contained ~2.0% mass of nickel oxide (calculated per metal Ni), the specific surface was 34 $m^2/g$, pore volume 0.03 $cm^3/g$, average pore diameter ~8 nm.[133] According to XRD, nickel oxide is present in NiO/GFC in the form of β-NiO phase (structural type NaCl, space group *Fm3m*); the average crystallite size calculated by Scherrer formula[107] is equal to 12–17 nm. The UV-Vis DR spectrum of the GFC sample with $Ni^{2+}$ cations contains the absorption bands of 13,800, 24,000, and 26200 $cm^{-1}$, attributed to d–d-transitions of $Ni^{2+}$ cations in the octahedral oxygen coordination ($Ni^{2+}_{Oh}$); these cations are stabilized in the structure of NiO.[135] As demonstrated by TEM images (Figure 2.17), nickel oxide in the form of large porous agglomerates (45–700 nm) with particle size of 15–25 nm is localized on the secondary support surface or apart from it. It is also possible to observe the smaller NiO particles with sizes of 3–5 nm inside the silica matrix.

The synthesized GFC was modified by the deposition of carbon nanofibers (CNF) afterwards, which are formed in the reaction of catalytic pyrolysis of light paraffins. Carbonization of the catalyst sample was performed in the metal tubular reactor at ambient pressure and at a temperature equal to 450°C. The GFC was loaded into the reactor in a form of the cylindrical structured cartridge with the catalytic fabric rolled into a spiral with a corrugated steel foil. LPG (propane-butane) was used as a carbonization reaction medium.

In the beginning of the carbon deposition process, a significant concentration of methane (up to 50 vol.%) and hydrogen (up to 40 vol.%) was observed in the outlet gases (Figure 2.18).

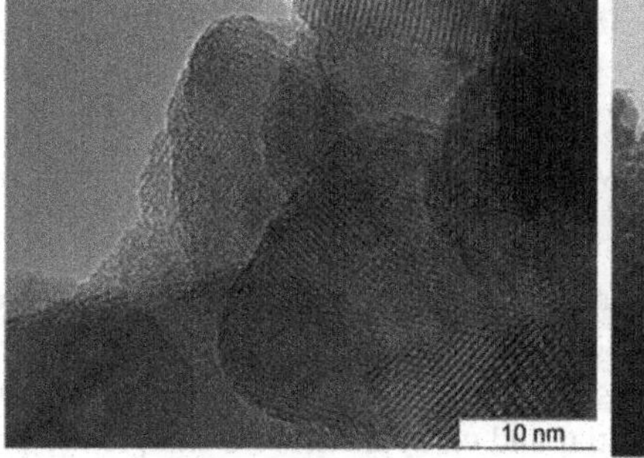

**FIGURE 2.17** TEM images of the initial NiO/GFC.

*Source:* Reprinted with permission from Popov M. V., et al. 2017.[133]

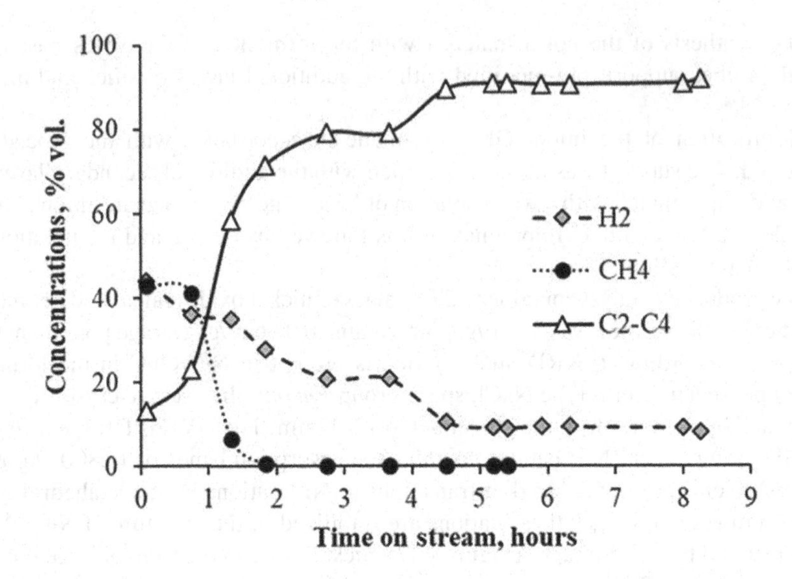


**FIGURE 2.18** Evolution of the volume concentration of propane-butane pyrolysis products at Ni/$SiO_2$/GFC in time.

*Source:* Reprinted with permission from Popov M. V., et al. 2017.[133]

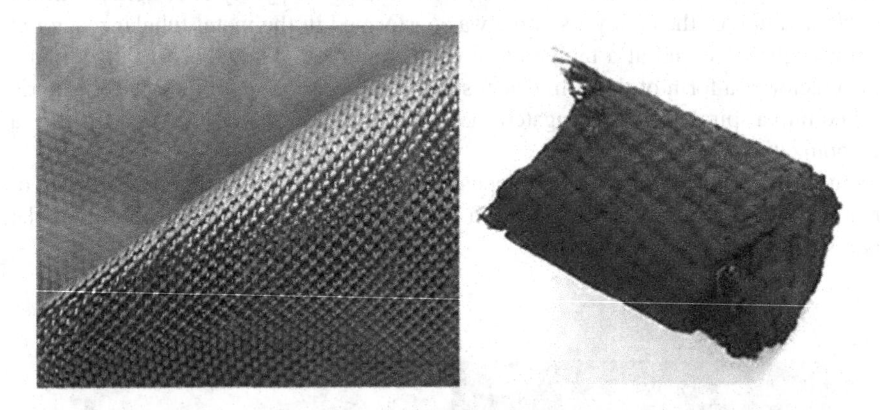

**FIGURE 2.19** View of the initial (on the left) and carbonized (on the right) Ni/GFC.

*Source:* Reprinted with permission from Popov M. V., et al. 2017.[133]

During the first hour, the LPG conversion reached 90%. Later, the formation of the gaseous pyrolysis products decreased, and emission of methane completely stopped in 1.5 hours, though hydrogen was still presented in the outlet gases during next 6 hours. The external view of the initial and carbonized GFC sample are presented in Figure 2.19.

As shown by thermo-gravimetry and differential scanning calorimetry performed in air medium (Figure 2.20), the CNF oxidation by oxygen in the carbonized GFC

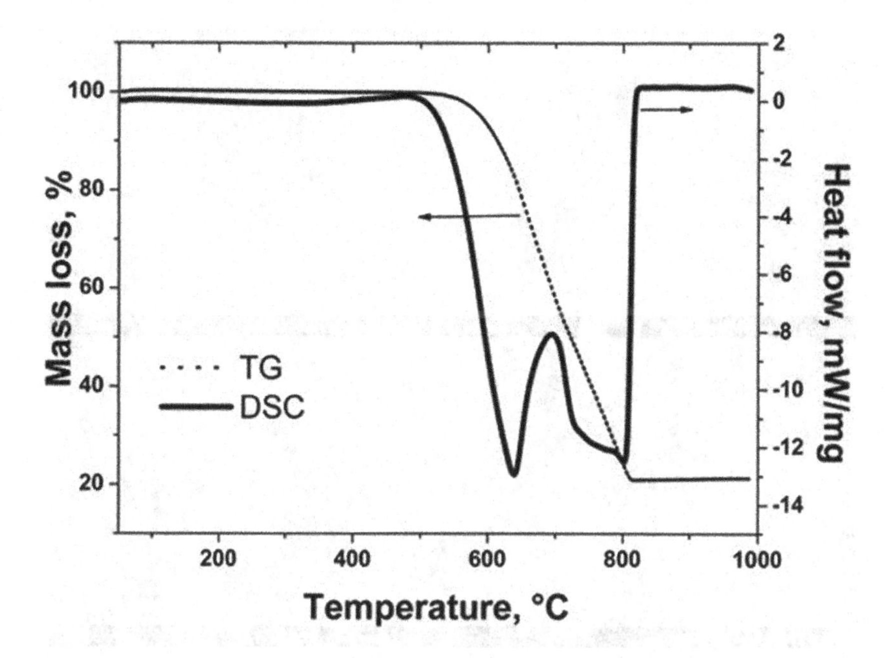


**FIGURE 2.20** Thermo-gravimetry and differential scanning calorimetry plot.

sample starts at ~493°C and finishes at 812°C. The temperature of the oxidation start is rather low; this may be attributed to the high disorder level of the carbon material, what is typical for CNFs.

Data on the porous structure of the carbonized GFC demonstrated that the total pore volume is equal to 0.18 $cm^3/g$, the specific surface area ~100 $m^2/g$, twice exceeding that for the GFC sample with alumina secondary support.[94]

SEM images of the carbonized sample (Figure 2.21) revealed that the deposited CNFs have the coaxial-conical packing of the graphite layers and represent themselves the nanofiber "forest," situated around the initial glass microfiber. Diameter of the nanofibers varies from 30 to 200 nm; they may reach few dozens of microns in length. The CNFs from the neighboring glass fibers may intertwist with each other, thus providing the high mechanical strength of the carbonized material.

The proposed method of the synthesis[134] is applicable for the production of the carbonized glass-fiber–based material with the high content of CNFs, improved mechanical stability, strong bonding of CNF to the support surface, and developed specific surface area. The method is technologically simple, easily scalable, and cost-efficient, thus providing the promising competitive characteristics of the produced material.

The high value of the surface area of the synthesized material makes it an attractive support for various catalysts, appropriate for further supporting significant amounts of active components, thus expanding the area of the practical applications of GFCs. Hydrophobic properties of the material can be also useful for the catalytic applications in the gaseous humid medium.

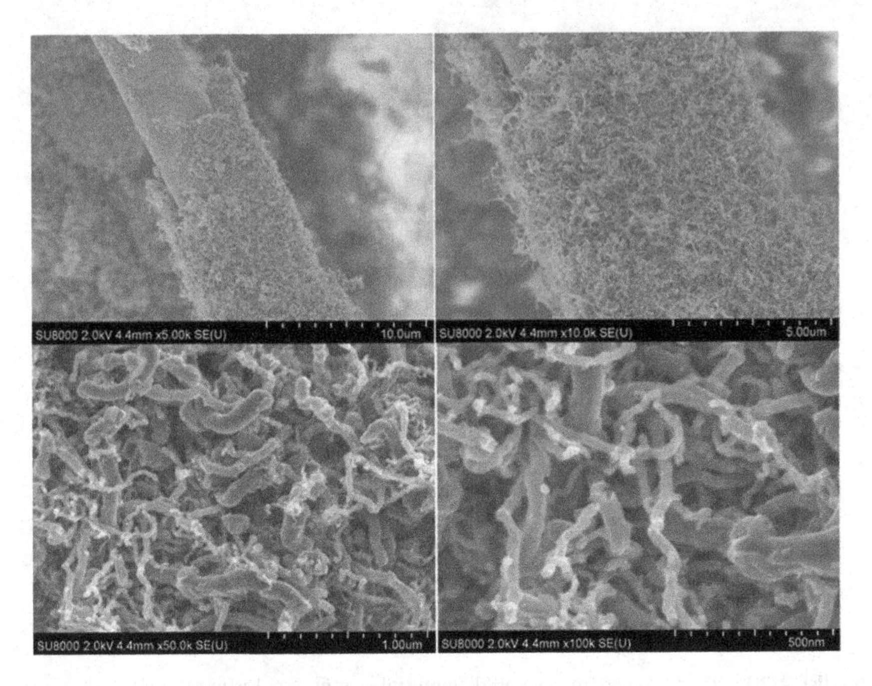


**FIGURE 2.21** SEM images of CNFs deposited at Ni/GFC surface.

*Source:* Reprinted with permission from Popov M. V., et al. 2017.[133]

The manufactured carbonized material is very solid and mechanically strong, and thus able to keep the preassigned shape, which can be easily varied as the initial GFC template has excellent flexibility. This makes it possible to produce the structured catalytic cartridges with the required external shape and internal geometric structure for the provision of the intensive mass transfer and low hydraulic resistance. High mechanical strength of the synthesized material opens the way to production of such cartridges with a minimum use of the steel flat or corrugated wire meshes and other additional structuring elements or even without using them at all.

In general, the described material offers great opportunities from the catalytic and engineering points of view.

## 2.7. CONCLUSION

We can conclude that the microfibrous catalysts are very promising for the application in different catalytic reactions. The method of surface thermal synthesis makes it possible to produce highly dispersed particles of the active component at the fiber surface. Modification of the fibers with the support of the secondary porous layer allows for the deposit of significant amounts of the active components (up to 10 mass % and more). The technology is based on the application of inexpensive and widely available types of a glass-fiber fabric; it is simple, easily scalable, and practically free of any manufacturing wastes.

In general, this synthesis approach opens the way for the production of very different GFCs for numerous applications.

However, it is doubtful that GFC may contain any unique active catalytic sites, which are really different from those at conventional supports. In our opinion, the unique properties and practical potential of the considered GFCs is mostly in their unusual geometric structure and mechanical flexibility. The latter is especially interesting, as no flexible catalysts have been ever put into practice before. All these properties allow for the design of novel catalytic beds with increased efficiency of heat and mass transfer and low pressure drop, thus opening new paths for creative chemical engineers.

## REFERENCES


1. Diemel F. 1932. British Patent No.364342. An improved catalytic unit for heating.
2. Bertsch J.A. 1944. US Patent No.2349844. Catalyst carrier.
3. Clare Brothers Ltd. 1965. British Patent No.1014846. Radiant burner.
4. Hauel A.P. 1965. US Patent No.3189563. Open mesh glass fabric supported catalyst.
5. Elmer T.H., Tischer R.E. 1974. US Patent No.3804647. Porous glass supports for automotive emissions control catalysts.
6. Nakamura H. 1974. British Patent No.1372806. Exhaust gas purifying apparatus.
7. Nakamura H. 1975. US Patent No.3897366. Automotive exhaust gas catalyst.
8. Nicholas D.M., Shah Y.T. 1976. Carbon monoxide oxidation over a platinum-porous fiber glass supported catalyst. *Ind Eng Chem Prod RD.* 15: 35–40.
9. Weber G.F., Ness S.R., Laudal D.L. 1991. Simultaneous NOx and particulate control using a catalyst-coated fabric filter. International Joint Power Generation Conference (San Diego, CA, October 6–10, ASME Technical Paper N 91-JPGC-FACT-2).
10. Barelko V.V., Khalzov P.I., Zviagin V.N. et al. 1996. Russian Patent No.2069584. Catalyst for chemical processes, namely, for conversion of ammonia, oxidation of hydrocarbons and sulfur dioxide, purification of exhaust gases.
11. Barelko V.V., Yuranov I.A., Cherashev A.F. et al. 1998. NO reduction on the catalytic-systems on the base of amorphous matrices with addition of metals or their oxides. *Dokl. Akad. Nauk.* 361: 485–8.
12. Kiwi-Minsker L., Yuranov I., Siebenhaar B. et al. 1999. Glass fiber catalysts for total oxidation of CO and hydrocarbons in waste gases. *Catal Today.* 54(1): 39–46.
13. Barelko V.V., Balzhinimaev B.S., Kildyashev S.P. et al. 2000. Russian Patent No.2143948. Support and catalyst for heterogeneous reactions.
14. Simonova L.G., Barelko V.V., Lapina O.B. et al. 2001. Catalysts based on fiberglass supports: I. Physicochemical properties of silica fiberglass supports. *Kinet Catal.* 42: 693–702.
15. Simonova L.G., Barelko V.V., Paukshtis E.A. et al. 2001. Catalysts based on fiberglass supports: II. Physicochemical properties of alumina borosilicate fiberglass supports. *Kinet Catal.* 42: 828–36.
16. Simonova L.G., Barelko V.V., Toktarev A.V. et al. 2001. Catalysts based on fiberglass supports: III. Properties of supported metals (Pt and Pd) according to electron-microscopic and XPS data. *Kinet Catal.* 42: 837–46.
17. Simonova L.G., Barelko V.V., Toktarev A.V. et al. 2002. Catalysts based on fiberglass support: IV. Platinum catalysts based on fiberglass support in oxidation of hydrocarbons (Propane and n-Butane) and sulfur dioxide. *Kinet Catal.* 43: 61–6.
18. Balzhinimaev B.S., Barelko V.V., Suknev A.P. et al. 2002. Catalysts based on fiberglass supports: V. Absorption and catalytic properties of palladium catalysts based on a

leached silica-fiberglass support in the selective hydrogenation of an ethylene-acetylene mixture. *Kinet Catal.* 43: 542–9.
19. Balzhinimaev B.S., Simonova L.G., Barelko V.V. et al. 2003. Pt-containing catalysts on a base of woven glass fiber support: A new alternative for traditional vanadium catalysts in $SO_2$ oxidation process. *Chem Eng J.* 91(2–3): 175–9.
20. Balzhinimaev B.S., Paukshtis E.A., Simonova L.G. et al. 2004. Oxidative destruction of chloroorganic compounds at glass-fiber catalysts. *Catalysis in Industry.* 5: 21–27.
21. Balzhinimaev B.S., Paukshtis E.A., Zagoruiko A.N. et al. 2005. Russian Patent No.2257952. Catalytic system for heterogeneous reactions.
22. Balzhinimaev B.S., Paukshtis E.A., Zagoruiko A.N. et al. 2005. Russian Patent No.2250890. Method for catalytic chlorination of lower alkanes with production of valuable products.
23. Balzhinimaev B.S., Paukshtis E.A., Zagoruiko A.N. et al. 2005. Russian Patent No.2250891. Method for production of vinyl chloride.
24. Balzhinimaev B.S., Paukshtis E.A., Zagoruiko A.N. et al. 2005. Russian Patent No.2252915. Method for sulfur dioxide oxidation.
25. Balzhinimaev B.S., Paukshtis E.A., Zagoruiko A.N. et al. 2005. Russian Patent No.2252208. Method for utilization of chloroorganic compounds.
26. Zagoruiko A.N., Veniaminov S.A., Veniaminova I.N. et al. 2007. Kinetic instabilities and intra-thread diffusion limitations in CO oxidation reaction at Pt/fiber-glass catalysts. *Chem Eng J.* 134: 111–16.
27. Zagoruiko A.N., Balzhinimaev B.S. 2011. Catalytic processes on the base of glass-fiber catalysts. *Chemical Industry Today.* 2: 5–11.
28. Balzhinimaev B.S., Paukshtis E.A., Lapina O.B. et al. 2010. Glass fiber materials as a new generation of structured catalysts. *Stud Surf Sci Catal.* 175: 43–50.
29. Balzhinimaev B.S., Paukshtis E.A., Vanag S.V. et al. 2010. Glass-fiber catalysts: Novel oxidation catalysts, catalytic technologies for environmental protection. *Catal Today.* 151(1–2): 195–9.
30. Balzhinimaev B.S., Parmon V.N. 2012. The innovative Russian approaches to catalysts design: New generation of fiberglass catalysts. *Top Catal.* 55(19–20): 1289–96.
31. Zagoruiko A.N., Lopatin S.A., Balzhinimaev B.S. et al. 2010. The process for catalytic incineration of waste gas on IC-12-S102 platinum glass fiber catalyst. *Catalysis in Industry.* 2: 113–17.
32. Zagoruiko A.N., Balzhinimaev B.S., Lopatin S.A. et al. 2010. Commercial process for incineration of VOC in waste gases on the base fiber-glass catalyst. Proceedings of XIX International Conference on Chemical Reactors CHEMREACTOR-19 (Vienna, Austria, September 5–9): 586–7.
33. Zagoruiko A., Balzhinimaev B., Vanag S. et al. 2010. Novel catalytic process for flue gas conditioning in electrostatic precipitators of coal-fired power plants. *J Air Waste Manag Assoc.* 60: 1002–8.
34. Paukshtis E.A., Simonova L.G., Zagoruiko A.N. et al. 2010. Oxidative destruction of chlorinated hydrocarbons on Pt-containing fiber-glass catalysts. *Chemosphere.* 79(2): 199–204.
35. Petrov S.V. 2000. Development of fibrous catalysts for oxidation of sulphide-containing substances in industrial wastes. PhD Thesis. Saint-Petersburg State University of Technology and Design.
36. Vitkovskaya R.F. 2005. Development and investigation of fibrous catalysts and contact elements for resource saving and environmental protection. DrSc Thesis. Saint-Petersburg State University of Technology and Design.
37. Matatov-Meytal Y., Sheintuch M. 2002. Catalytic fibers and cloths. *Appl Catal A-Gen.* 231(1–2): 1–16.

38. Kotolevich Y.S. 2012. Thermal synthesis of silver catalysts supported at glass-fiber fabric. PhD Thesis. Omsk: Institute for Hydrocarbon Processing.
39. Reichelt E., Heddrich M.P., Jahn M. et al. 2014. Fiber based structured materials for catalytic applications. *Appl Catal A-Gen.* 476: 78–90.
40. Tianjie Pei, Laishuan Liu, Longkun Xu et al. 2016. A novel glass fiber catalyst for the catalytic combustion of ethyl acetate. *Catal Commun.* 74: 19–23.
41. Krauns C., Barelko V., Fabre G. et al. 2001. Fiber glass supported catalysts and pure platinum: Laser ignition of catalytic combustion of propane. *Catal Lett.* 72(3–4): 161–5.
42. Medina-Valtierra J., Ramírez-Ortiz J., Arroyo-Rojas V.M. et al. 2003. Cyclohexane oxidation over Cu2O—CuO and CuO thin films deposited by CVD process on fiberglass. *Appl Catal A-Gen.* 238: 1–9.
43. Brüning R., Scholz P., Morgenthal I. et al. 2005. Innovative catalysts for oxidative dehydrogenation in the gas phase—metallic short fibers and coated glass fabrics. *Chem Eng Technol.* 28(9): 1056–62.
44. Zazhigalov S., Elyshev A., Lopatin S. et al. 2017. Copper-chromite glass fiber catalyst and its performance in the test reaction of deep oxidation of toluene in air. *React Kinet Mech Cat.* 120(1): 247–60.
45. Balzhinimaev B.S., Paukshtis E.A., Zagoruiko A.N. et al. 2007. Russian Patent No.2305090. Method for deep oxidation of light paraffins.
46. Lopatin S.A., Tsyrul'nikov P.G., Kotolevich Y.S. et al. 2015. Structured woven glass-fiber IC-12-S111 catalyst for the deep oxidation of organic compounds. *Catalysis in Industry.* 7(4): 329–34.
47. Matatov-Meytal, Y., Barelko, V., Yuranov, I. et al. 2000. Cloth catalysts in water denitrification: I. Pd on glass fibers, *Appl Catal B-Environ.* 27(2): 127–35.
48. Matatov-Meytal, Y., Barelko, V., Yuranov, I. et al. 2001. Cloth catalysts for water denitrification: II. Removal of nitrates using Pd—Cu supported on glass fibers. *Appl Catal B-Environ.* 31(4): 233–40.
49. Zagoruiko A., Vanag S., Balzhinimaev B. et al. 2009. Catalytic flue gas conditioning in electrostatic precipitators of coal-fired power plants. *Chem Eng J.* 154: 325–32.
50. Vanag S.V., Zagoruiko A.N., Zykov A.M. 2012. Chemisorption and oxidation of $SO_2$ at Pt-containing fiber-glass catalysts. Proceedings of IX International Conference on Mechanisms of Catalytic Reactions (St. Petersburg, October 22–25): 272.
51. Vanag S.V. 2012. Processes for oxidation of $SO_2$ into $SO_3$ using glass-fiber catalysts and their equipment arrangement. PhD Thesis. Novosibirsk: Boreskov Institute of Catalysis.
52. Vanag S.V., Paukshtis E.A., Zagoruiko A.N. 2015. Properties of platinum-containing glass-fiber catalysts in the $SO_2$ oxidation reaction. *React Kinet Mech Cat.* 116: 147–58.
53. Zagoruiko A.N., Glotov V.D., Lopatin S.A. et al. 2016. Investigation of the internal structure, fluid flow dynamics and mass transfer in the multi-layered packing of glass-fiber catalyst in the pilot reactor for sulfur dioxide oxidation. *Science Bulletin of The Novosibirsk State Technical University.* 3(64): 161–77.
54. Zagoruiko A.N., Shinkarev V.V., Simonova L.G. 2007. Vanadia/glass-fiber catalyst for hydrogen sulfide selective oxidation to sulfur by oxygen. Proceedings of 3rd International Conference "Catalysis: Fundamentals and Application (Novosibirsk, Russia, July 4–8): 2, 580.
55. Mikenin P.E., Lopatin S.A., Zazhigalov S.V. et al. 2015. Structured glass-fiber catalysts for selective oxidation of $H_2S$. Proceedings of 12th European Congress on Catalysis—EuropaCat-XII (Kazan, Russia, August 30–September 4): 1813–14.
56. Mikenin P.E., Tsyrul'nikov P.G., Kotolevich Y.S. et al. 2015. Vanadium oxide catalysts on structured microfiber supports for the selective oxidation of hydrogen sulfide. *Catalysis in Industry.* 7(2): 155–60.

57. Mikenin P., Zazhigalov S., Elyshev A. et al. 2016. Iron oxide catalyst at the modified glass fiber support for selective oxidation of $H_2S$. *Catal Commun.* 87: 36–40.
58. Larina T.V., Cherepanova S.V., Rudina N.A. et al. 2015. Characterization of vanadia catalysts on structured micro-fibrous glass supports for selective oxidation of hydrogen sulfide. *Catalysis for Sustainable Energy.* 2: 87–95.
59. Elyshev A., Larina T., Cherepanova S. et al. 2016. Physical and chemical properties of FeOx-based glass fiber catalyst synthesized by surface thermo-synthesis method. Proceedings of II Scientific Technological Symposium "Catalytic Hydroprocessing in Oil Refining" (Belgrade, Serbia, April 17–23): 115–16.
60. Aldashukurova G.B., Mironenko A.V., Mansurov Z.A. et al. 2013. Synthesis gas production on glass cloth catalysts modified by Ni and Co oxides. *J Energy Chem.* 22(5): 811–18.
61. Britcher L.G., Matisons J.G. 2000. E-glass fiber supported hydrosilation catalysts. *ACS Symposium Series.* 760: 127–44.
62. Li L., Diao Y., Liu X. 2014. Ce-Mn mixed oxides supported on glass-fiber for low-temperature selective catalytic reduction of NO with $NH_3$. *J Rare Earth.* 32(5): 409–15.
63. Balzhinimaev B.S., Paukshtis E.A., Zagoruiko A.N. et al. 2008. Russian Patent No.№2330834. Method for selective catalytic methane chlorination into methyl chloride.
64. Shalygin A., Paukshtis E., Kovalyov E. et al. 2013. Light olefins synthesis from C1-C2 paraffins via oxychlorination processes. *Front Chem Sci Eng.* 7(3): 279–88.
65. Debeche T., Marmet C., Kiwi-Minsker L. et al. 2005. Structured fiber supports for gas phase biocatalysis. *Enzyme Microb Tech.* 36(7): 911–16.
66. Hofstadler K., Bauer R., Novalic S. et al. 1994. New reactor design for photocatalytic wastewater treatment with TiO2 immobilized on fused-silica glass fibers: Photomineralization of 4-chlorophenol. *Environ Sci Technol.* 28(4): 670–4.
67. Palau J., Colomer M., Penya-Roja J.M. et al. 2012. Photodegradation of toluene, m-xylene, and n-butyl acetate and their mixtures over $TiO_2$ catalyst on glass fibers. *Ind Eng Chem Res.* 5(17): 5986–94.
68. Sangkhun W., Laokiat L., Tanboonchuy V. et al. 2012. Photocatalytic degradation of BTEX using W-doped $TiO_2$ immobilized on fiberglass cloth under visible light. *Superlattice Microst.* 52(4): 632–42.
69. Lin S., Zhang X., Sun Q. et al. 2013. Fabrication of solar light induced Fe-$TiO_2$ immobilized on glass-fiber and application for phenol photocatalytic degradation. *Mater Res Bull.* 48(11): 4570–5.
70. Pham T-D., Lee B-K. 2014. Feasibility of silver doped $TiO_2$/glass fiber photocatalyst under visible irradiation as an indoor air germicide. *Int J Env Res Pub He.* 11(3): 3271–88.
71. Salmi T., Mäki-Arvela P., Toukoniitty E. et al. 2000. Immobile silica fibre catalyst in liquid-phase hydrogenation. *Stud Surf Sci Catal.* 130: 2033–8.
72. Balzhinimaev B.S., Paukshtis E.A., Zagoruiko A.N. et al. 2006. Russian Patent No.2289565. Method for selective hydrogenation of acetylene compounds in olefin-rich media.
73. Zagoruiko A.N., Arendarskii D.A., Balzhinimaev B.S. 2007. Russian Patent for Utility Model No.66976A. Catalytic system for selective hydrogenation of acetylenes in the media of olefins and diolefins.
74. Balzhinimaev B.S., Zagoruiko A.N., Gilmutdinov N.R. et al. 2008. Russian Patent for Utility Model No.79054. Catalytic system for purification of diolefins from admixtures of acetylene compounds.
75. Gulyaeva Y.K., Kaichev V.V., Zaikovskii V.I. et al. 2015. Selective hydrogenation of acetylene over novel Pd/fiberglass catalysts. *Catal Today.* 245: 139–46.
76. Balzhinimaev B.S., Paukshtis E.A., Zagoruiko A.N. et al. 2007. Russian Patent No.2292950. Catalytic system for heterogeneous reactions.

77. Xanthopoulou G., Vekinis G. 1998. Investigation of catalytic oxidation of carbon monoxide over a Cu—Cr-oxide catalyst made by self-propagating high-temperature synthesis. *Appl Catal B-Environ.* 19(1): 37–44.
78. Specchia S., Civera A., Saracco G. 2004. In situ combustion synthesis of perovskite catalysts for efficient and clean methane premixed metal burners. *Chem Eng Sci.* 59: 5091–8.
79. Zav'yalova U.F., Tret'yakov V.F., Burdeinaya T.N. et al. 2005. Self-propagating synthesis of supported oxide catalysts for deep oxidation of CO and hydrocarbons. *Kinet Catal.* 46(5): 752–7.
80. Fino D., Russo N., Saracco G. et al. 2006. Catalytic removal of NOx and diesel soot over nanostructured spinel-type oxides. *J Catal.* 242: 38–47.
81. Mukasyan A.S., Epstein P., Dinka P. 2007. Solution combustion synthesis of nanomaterials. *P Combust Inst.* 31: 1789–95.
82. Kumar A., Mukasyan A.S., Wolf E.E. 2011. Combustion synthesis of Ni, Fe and Cu multi-component catalysts for hydrogen production from ethanol reforming. *Appl Catal A-Gen.* 401: 20–8.
83. Morsi K. 2012. The diversity of combustion synthesis processing: A review. *J Mater Sci.* 47: 68–92.
84. Yadav G.D., Ajgaonkar N.P., Varma A. 2012. Preparation of highly superacidic sulfated zirconia via combustion synthesis and its application in Pechmann condensation of resorcinol with ethyl acetoacetate. *J Catal.* 292: 99–10.
85. Postole G., Nguyen T-S., Aouine M. et al. 2015. Efficient hydrogen production from methane over iridium-doped ceria catalysts synthesized by solution combustion. *Appl Catal B-Environ.* 166–167: 580–91.
86. Desyatikh I.V., Vedyagin A.A., Kotolevich Y.S. et al. 2011. Preparation of CuO-CeO2 catalysts deposited on glass cloth by surface self-propagating thermal synthesis. *Combust Explo Shock.* 47: 677–82.
87. Afonasenko T.N., Tsyrul'nikov P.G., Gulyaeva T.I. et al. 2013. (CuO-$CeO_2$)/glass cloth catalysts for selective CO oxidation in the presence of $H_2$: The effect of the nature of the fuel component used in their surface self-propagating high-temperature synthesis on their properties. *Kinet Catal.* 54: 59–68.
88. Mironenko O.O., Shitova N.B., Kotolevich Y.S. et al. 2012. Pd/fiber glass and Pd/5% γ-Al2O3/fiber glass catalysts by surface self-propagating thermal synthesis. *International Journal of Self-Propagating High-Temperature Synthesis.* 21: 139–45.
89. Kotolevich Y.S., Khramov E.V., Mironenko O.O. et al. 2014. Supported palladium catalysts prepared by surface self-propagating thermal synthesis. *International Journal of Self-Propagating High-Temperature Synthesis.* 23: 9–17.
90. Kotolevich Y.S., Sigaeva S.S., Tsyrulnikov P.G. et al. 2015. Russian Patent No.2549906. Method for preparation of supported catalysts by means of impulse surface thermal synthesis.
91. Kulikov A.V., Zagoruiko A.N., Lopatin S.A. et al. 2015. Catalytic heating element based on the platinum-containing glass-fiber catalyst IC-12-S111. *Science Bulletin of the Novosibirsk State Technical University.* 1(58): 257–70.
92. Yagi J. 1978. Japan Patent Application JPS5363296. Production of catalyst carrier.
93. Yagi J., Nakamura H., Maezava Y. 1977. British Patent No.1460748. Catalyst for exhaust gas purification.
94. Ismagilov Z.R., Shikina N.V., Kruchinin V.N. et al. 2005. Development of methods of growing carbon nanofibers on silica glass fiber supports. *Catal Today.* 102–103: 85–93.
95. Brichkov A.S., Brichkova V.Y., Kozik V.V. et al. 2015. Russian Patent No.2538206. Method for production of catalyst for incineration of propane at glass-fiber support.
96. Elyshev A., Larina T., Cherepanova S. et al. 2016. Physical and chemical properties of CuCr2O4-based glass fiber catalyst synthesized by surface thermo-synthesis method.

Proceedings of II Scientific Technological Symposium "Catalytic Hydroprocessing In Oil Refining" (Belgrade, Serbia, April 17–23): 113–14.
97. Zhurba E.N., Lavrinovich I.A., Trofimov A.N. et al. 2001. Russian Patent No.2165393. Glass for manufacturing fiber glass and high- temperature silicic fiber glass based thereon.
98. Zagoruiko A.N., Lopatin S.A., Zazhigalov S.V. et al. 2017. Russian Patent No.2633369. Method for preparation of microfibrous catalyst.
99. Lopatin S., Chub O., Yazykov N. et al. 2014. Structured cartridges with reinforced fiber-glass catalyst for fuel combustion in the fluidized beds of the inert heat-transfer particles. Proceedings of the XXI International Conference on Chemical Reactors "CHEMREACTOR-21" (Delft, The Netherlands, September 22–25): 272–3.
100. Mulina T.V., Filipov A.V., Chumachenko V.A. 1988. CO oxidation over CuCr/Al2O3 catalysts in presence of SO2. *React Kinet Catal Lett.* 37: 95–100.
101. Hosseini S.A., Niaei A., Salari D. et al. 2014. Study of correlation between activity and structural properties of Cu-(Cr, Mn and Co)2 nano mixed oxides in VOC combustion. *Ceram Int.* 40: 6157–63.
102. Mazzocchia C., Kaddouri A. 2003. On the activity of copper chromite catalysts in ethyl acetate combustion in the presence and absence of oxygen. *J Mol Catal A-Chem.* 204–205: 647–54.
103. Carotenuto G., Kumar A., Miller J. et al. 2013. Hydrogen production by ethanol decomposition and partial oxidation over copper/copper-chromite based catalysts prepared by combustion synthesis. *Catal Today.* 203: 163–75.
104. Chiu T.W., Yu B.S., Wang Y.R. et al. 2011. Synthesis of nanosized $CuCrO_2$ porous powders via a self-combustion glycine nitrate process. *J Alloy Compd.* 509: 2933–5.
105. Shankha S.A. 2014. Preparation of the CuCr2O4 spinel nanoparticles catalyst for selective oxidation of toluene to benzaldehyde. *Green Chem.* 16: 2500–8.
106. Massa W. 2005. *Crystal structure determination*. Berlin: Springer.
107. David W.I.F., Shankland K., McCusker L.B. et al. 2002. *Structure determination form powder diffraction data.* Oxford: Oxford Science publications.
108. Plyasova L.M., Larina T.V., Krivencov V.V. et al. 2015. Effect of the Cr/Fe ratio on the structure of Fe-Cr-Cu-containing oxide catalysts. *Kinet Catal.* 56: 493–500.
109. Asmolov G.N., Krylov O.V. 1970. Investigation of molybdenum oxide catalysts, supported at $\alpha$-$Al_2O_3$ и MgO by means of diffusion reflectance spectra. *Kinet. Catal.* 11: 1028–33.
110. Lever A.B.P. 1987. *Inorganic electronic spectroscopy.* 2nd ed. Amsterdam: Elsevier.
111. Zagoruiko A.N., Lopatin S.A., Mikenin P.E. et al. 2017. Novel structured catalytic systems—cartridges on the base of fibrous catalysts. *Chem Eng Proc: Proc Int.* 122: 460–72.
112. Zagoruiko A.N., Shinkarev V.V., Vanag S. et al. 2010. Catalytic processes and catalysts for production of elemental sulfur from sulfur-containing gases. *Catalysis in Industry.* 2(4): 343–52.
113. Piéplu A., Saur O., Lavalley J.C. et al. 1998. Claus catalysis and H2S selective oxidation. *Cataly Rev.* 40: 409–50.
114. Ismagilov Z.R., Khairulin S.R., Kerzhentsev M.A. et al. 1999. Development of catalytic technologies for purification of gases from hydrogen sulfide based on direct selective catalytic oxidation of $H_2S$ to elemental sulfur. *Eurasian Chemico-Technological Journal.* 1: 49–56.
115. Marshneva V.I., Mokrinskii V.V. 1989. Catalytic activity of metal oxides in hydrogen sulfide oxidation by oxygen and sulfur dioxide. *Kinet Catal.* 29: 854–8.
116. Davydov A.A., Marshneva V.I., Shepotko M.L. 2003. Metal oxides in hydrogen sulfide oxidation by oxygen and sulfur dioxide: I. The comparison study of the catalytic activity. Mechanism of the interactions between $H_2S$ and $SO_2$ on some oxides. *Appl Catal A-Gen.* 244: 93.

117. Marusak L.A., Messier R., White W.B. 1980. Optical absorption spectrum of hematite, $\alpha$-$Fe_2O_3$ near IR to UV. *J Phys Chem Solid*. 41(9): 981–4.
118. Tran C., Singhal R., Lawrence D. et al. 2015. Polyaniline-coated freestanding porous carbon nanofibers as efficient hybrid electrodes for supercapacitors. *J Power Sources*. 293: 373–9.
119. Lai C-C., Lo C-T. 2015. Preparation of nanostructural carbon nanofibers and their electrochemical performance for supercapacitors. *Electrochim Acta*. 183: 85–93.
120. Ferguson A.J., Blackburn J.L., Kopidakis N. 2013. Fullerenes and carbon nanotubes as acceptor materials in organic photovoltaics. *Mater Lett*. 90: 115–25.
121. Peng S., Li L., Kong J. et al. 2016. Electrospun carbon nanofibers and their hybrid composites as advanced materials for energy conversion and storage. *Nano Energy*. 22: 361–95.
122. Bera R.K., Mhaisalkar S.G., Mandler D. et al. 2016. Formation and performance of highly absorbing solar thermal coating based on carbon nanotubes and boehmite. *Energ Convers Manage*. 120: 287–93.
123. Hanaei H., Assadi M., Saidur R. 2016. Highly efficient antireflective and self-cleaning coatings that incorporate carbon nanotubes (CNTs) into solar cells: A review. *Renew Sust Energ Rev*. 59: 620–35.
124. Macutkevic J., Kuzhir P., Seliuta D. et al. 2010. Dielectric properties of a novel high absorbing onion-like-carbon based polymer composite. *Diam Relat Mater*. 19: 91–9.
125. Mazov I.N., Ilinykh I.A., Kuznetsov V.L. et al. 2014. Thermal conductivity of polypropylene-based composites with multiwall carbon nanotubes with different diameter and morphology. *J Alloy Compd*. 586: 440–2.
126. Kuzhir P., Maksimenko S., Bychanok D. et al. 2009. Nano-scaled onion-like carbon: Prospective material for microwave coatings. *Metamaterials*. 3: 148–56.
127. Shinkarev V.V., Glushenkov A.M., Kuvshinov D.G. et al. 2009. New effective catalysts based on mesoporous nanofibrous carbon for selective oxidation of hydrogen sulfide. *Appl Catal B-Environ*. 85: 180–91.
128. Shinkarev V.V., Glushenkov A.M., Kuvshinov D.G. et al. 2010. Nanofibrous carbon with herringbone structure as an effective catalyst of the $H_2S$ selective oxidation. *Carbon*. 48: 2004–12.
129. Shinkarev V., Kuvshinov G., Zagoruiko A. 2018. Kinetics of $H_2S$ selective oxidation by oxygen at the carbon nanofibrous catalyst. *React Kinet Mech Cat*. 123: 625–39.
130. Kuvshinov G.G., Mogilnykh Y.I., Kuvshinov D.G. et al. 1999. Mechanism of porous filamentous carbon granule formation on catalytic hydrocarbon decomposition. *Carbon*. 37: 1239–46.
131. Popov M.V., Shinkarev V.V., Brezgin P.I. et al. 2013. Effect of pressure on the production of hydrogen and nanofilamentous carbon by the catalytic pyrolysis of methane on Ni-containing catalysts. *Kinet Catal*. 54: 481–6.
132. Popov M., Zazhigalov S., Mikenin P. et al. 2016. Modification of glass fiber catalyst surface by additional layers of silica and carbon nanofibers. Proceedings of XXII International Conference on Chemical Reactors “CHEMREACTOR-22” (London, UK, September 19–23): 317–18.
133. Popov M.V., Zazhigalov S.V., Larina T.V. et al. 2017. Glass fiber supports modified by layers of silica and carbon nanofibers. *Catalysis for Sustainable Energy*. 4: 1–6.
134. Zagoruiko A.N., Lopatin S.A., Zazhigalov S.V. et al. 2017. Russian Patent No.2624216. Microfibrous catalyst support and method for its production.
135. Becerra A.M., Castro-Luna A.E. 2005. An investigation on the presence of $NiAl_2O_4$ in a stable Ni on alumina catalyst for dry reforming. *J Chil Chem Soc*. 50: 465–9.



# 3 Arrangement of the Beds of the Glass-Fiber Catalysts

## 3.1. STRUCTURING OF THE MICROFIBROUS CATALYSTS

The main distinction of GFCs from catalysts of conventional shapes is their original geometric structure and mechanical flexibility. These peculiarities open a unique possibility for creation of novel types of catalytic structures with various shapes, potentially characterized with a low pressure drop and enhanced heat and mass transfer properties. In turn, this makes it possible to develop catalytic reactors of a principally new design.

Basic structural elements of GFCs are the elementary glass fibers with a diameter typically in range from 1 to 10 microns; these fibers are used as a support for the active component. Obviously, the microfibers themselves cannot be used in a large-scale commercial catalytic reactor; they must be arranged into some mechanically stable packing with a regular macrostructure prior to their practical application.

In particular, the GFC microfibers may be structured in the form of fibrous wool or pad. The advantage of such GFC supports is the absence of geometric macrostructures, like threads and bundles in woven and knitted systems, which may cause problems with the internal heat and mass transfer limitations. However, in most cases these materials do not have sufficient mechanical strength and stability, which may lead to their destruction and fiber entrainment by the reaction flow. Besides, the results of the experimental studies of GFC-wool or GFC-pads[1] show that their internal structure is rather non-uniform and that quite significant non-uniformity of the reaction fluid distribution may exist along the catalyst bed volume.

The innovative technology used in the production of glass-fiber pads is worth mentioning; it is characterized by high strength and volume uniformity, as well as the possibility to include different types of fibers (such as glass fibers, metal fibers, etc.) and even microparticles of the granular catalysts into the pad volume.[2] However, this research is in the initial stage and mostly belongs to a future generation of GFCs.

Today, the most preferred form of the catalytic microfibers packing is a two-level system. At the first structural level GFC microfibers are twisted into threads (Figure 3.1), and at the second one these threads are arranged into a woven or knitted fabric either directly (Figure 3.2a) or through the preliminary arrangement of threads into bundles (Figure 3.2b).

Such materials are mechanically stable and have a high mechanical strength in combination with flexibility. These fabrics are manufactured in a wide variety of types in the glass industry worldwide; as a result, they are an inexpensive and widely available material for catalyst production.

**FIGURE 3.1** Structure of the microfibrous thread.

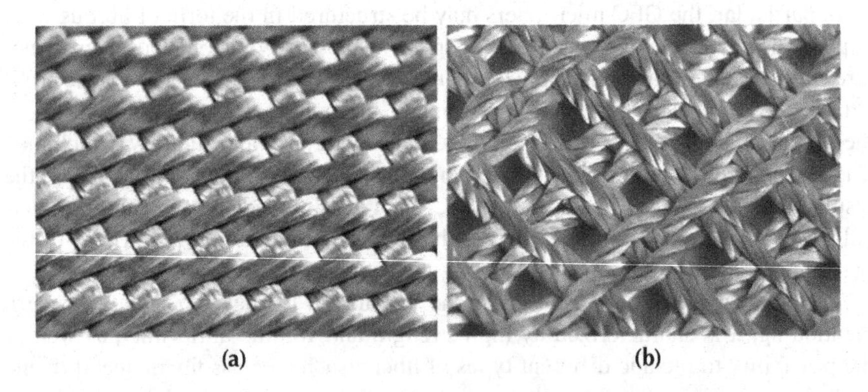

(a) (b)

**FIGURE 3.2** Images of GFC fabrics of different structure: sateen (a) and openwork (b) types.

At the next structural level these fabrics can be used to produce different packing for application in the catalytic reactors. In general, all methods of GFC fabric arrangement in such packings can be divided into two main groups, which differ in the orientation of the fabric plane in regard to the reaction fluid flow direction: systems with gliding (Figure 3.3a) and propagative (Figure 3.3b) flows.

Note that the different types of GFC packing provide the realization of various hydrodynamic regimes of reactor operation. For example, beds with gliding flow

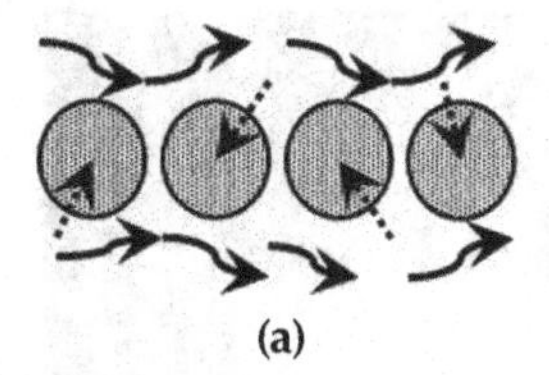
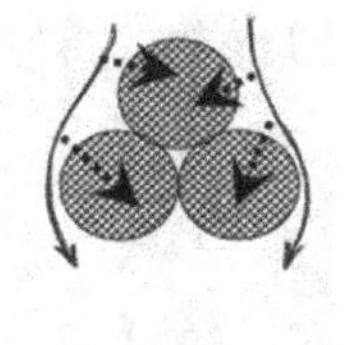
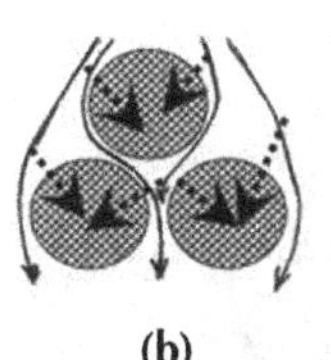
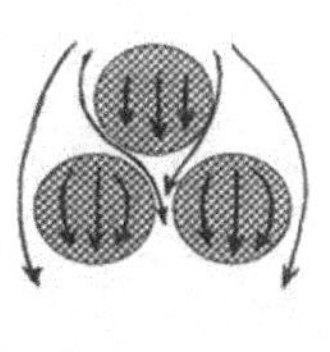

(a) (b)

**FIGURE 3.3** Fluid flow structure for gliding (a) and propagative (b) reaction mixture flow in relation to GFC threads. Solid arrows show convective flow; dashed arrows show diffusion flow. The cross-sections of the threads are shown with the shaded circles.

*Source:* Reprinted with permission from Lopatin et al. 2014.[3]

will more probably approach the plug flow regime, while in the case of propagative flow the continuous stirred-tank reactor (CSTR) regime will be more probable. Therefore, the selection of the optimal type of packing should include the consideration of the best possible regime for each given case. This factor should be also considered in the course of process optimization in case of reactions, characterized by kinetic instabilities and steady-state multiplicity. In particular, it was shown[4–6] that in case of carbon monoxide oxidation in thin beds of Pt-containing GFCs, the mutual transitions between plug flow and CSTR regimes may occur in addition to the known types of oscillations, hysteresis, and light-off behavior typical for CSTRs.

## 3.2. GFC PACKING WITH THE PROPAGATIVE FLOW OF REACTION FLUID

Depending upon the fluid movement velocity and its properties, the regimes with the intensive forced convective flow may exist inside the GFC threads in the propagative flow mode of GFC arrangement. Such regimes are interesting and promising for the practical application, as the convective intra-thread mass transfer is much more efficient than the diffusion-based one inside the catalysts of the conventional shapes (pellets, monoliths). Theoretically, in this case we may also expect the change of the external circumfluous surface of threads and fibers on the interaction between the GFC and fluid, leading, in some range of fluid velocities, to a high formal power order of the mass transfer coefficient dependence upon the fluid velocity (up to 1 and even higher).

The simplest variant of propagative GFC packing is a multilayer loading (Figure 3.4). This loading method, already known for more than 40 years,[7] was used, in particular, in the sulfur dioxide oxidation reactor at Byisk Oleum Plant.[8] It is quite trivial and intuitively understandable; however, it has definite disadvantages: rather high pressure drop, significant losses of GFC material during cutting of the catalyst circles for the round reactor from usually rectangle-shaped glass cloth, and limitation of the reactor size by the cloth width.

Another possible method of GFC propagative packing is zig-zag[9] or diagonal[10] beds (see Figure 3.5)

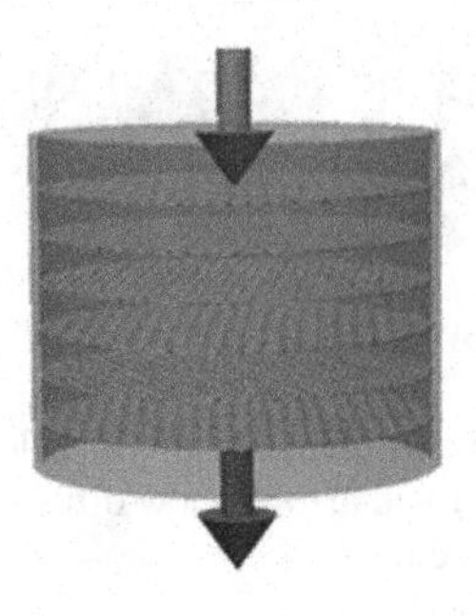

**FIGURE 3.4** Multilayered GFC packing with the propagative flow (flow direction is shown by arrows).

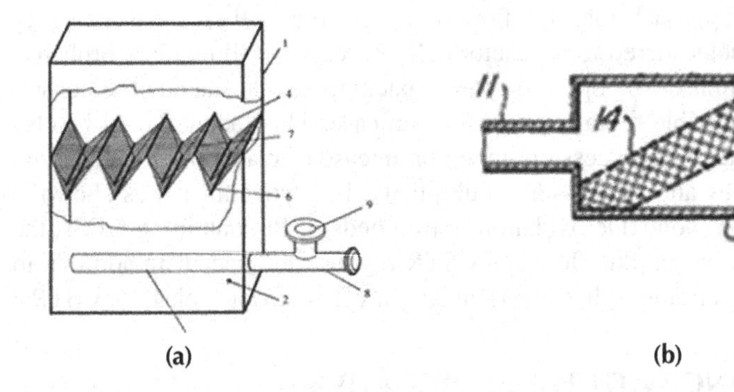


**FIGURE 3.5** Zig-zag (a) and diagonal (b) GFC packing.

Such arrangement of GFC layers allows for a decrease in reactor size and its pressure drop. On the other hand, such packing may raise the problem of the sealing of the GFC fabric edges. In addition, the maximum reactor width is limited by the cloth width, complicating the scale-up of such reactors.

The radial bed GFC packing[10–12] seems more advanced (Figure 3.6).

In this case it is possible to create the packing with a higher external area and reasonable amount of GFC layers, thus providing the low or moderate pressure drop even at high velocities of the reacting flow. Radial GFC beds were used in the reactor for the oxidation of chloroorganic compounds,[13] purification of diesel exhaust gases,[14] and selective hydrogenation of acetylenes in butadiene medium.[15]

The possible drawback of the radial beds is the non-uniform distribution of the reaction mixture flow along the packing height, though to some extent it is possible to minimize it by the modification of GFC bed internal design.[16]

High filtration ability in respect to solid and droplet particulate admixtures is a common feature for all variants of GFC beds with the propagative flow. Such admixtures can be observed in the gaseous or multiphase reacting medium, at that the consequences are rather doubtful. In some cases, it will be a serious advantage; for example, in the processes of catalytic incineration in gaseous streams, containing

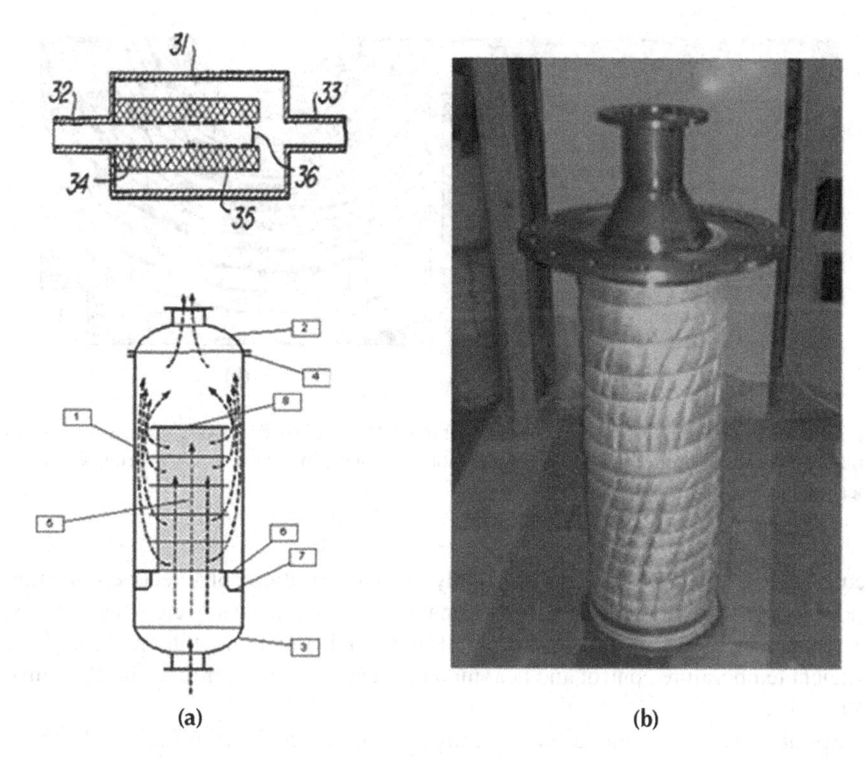


FIGURE 3.6 Radial beds of GFC (a) and view of the radial-type GFC reactor insert (b).

solid and liquid admixtures which may be oxidized, the filtration capacity of GFC layer will help to improve the abatement of these particulates. On the other hand, in case of dusty fluids processing, it can be a significant drawback, resulting in GFC bed clogging.

## 3.3. GFC PACKING WITH THE GLIDING FLOW OR REACTION MEDIA

### 3.3.1. CYLINDRICAL CARTRIDGES

It is possible to use the cylinder-shaped spiral cartridge[17–20] (see Figure 3.7) to ensure the gliding flow of the reaction mixture along the GFC fabric surface. Such cartridge represents a roulette, twisted from GFC fabric and structured gauze, necessary to form the regular system of numerous transport channels for the movement of the reaction media. The corrugated metal gauze (Figure 3.7a) or volumetric knitted net (Figure 3.7b) can play a role of such structuring element.

The spiral beds are characterized by a low pressure drop; they are also potentially applicable in the treatment of the heavily dusty flows.

Among promising future designs of spiral cartridges, it is worth mentioning the tubular reactor with spiral intra-tune GFC inserts[21] for the performance of reactions

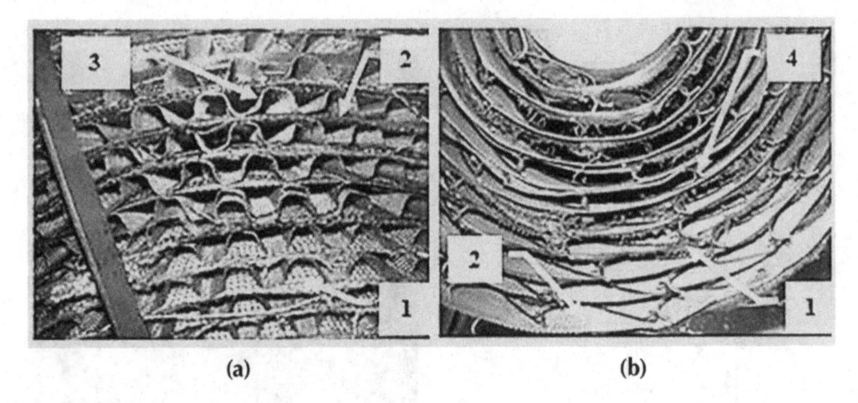


**FIGURE 3.7** Internal structure of the spiral catalytic cartridge using corrugated mesh (a) and volumetric knitted net (b): (1) GFC cloth, (2) plain metal mesh, (3) corrugated metal mesh, volumetric metal gauze.

requiring the intensive removal or supply of heat (reactions of selective oxidation and selective oxychlorination, dehydrogenation, steam conversion, etc.). Such inserts increase heat conductivity both in radial and axial directions, thus providing the efficient temperature control and heat management in the reactor tubes in the course of reaction.

Spiral blocks are suitable for charging into the reactors with a relatively small diameter. However, manufacturing, transporting, and charging the large spiral cartridges into the reactor becomes a seriously complicated task in case of larger diameters.

This problem may be solved by the application of a module approach, where the large catalyst beds are formed from the catalytic cartridges of a reasonable size and weight. Figure 3.8 shows the example of such modular bed.

Such an approach makes it possible to create beds of any size and shape. As such, the size of each cartridge is defined by the convenience of the charging and backward discharging from reactor. For instance, if the catalyst loading is performed through the reactor lid, then the cartridge diameter should be smaller than the lid diameter: the weight of each cartridge should be appropriate for manipulation during loading. However, as seen from Figure 3.8, in such a bed we will be inevitably confronted with the presence of void spaces between cartridges, as well as between the cartridges and reactor walls. Even though the fraction of such voids is decreasing with the increase of the reactor diameter $D$ to cartridge diameter $d$, this value cannot be smaller than a definite limiting value, which is 21.5% for "corridor" and 9.4% for "chess" placement of cartridges even for infinitely large $D/d$ values.

To exclude the reacting fluid bypass of catalyst cartridges, these voids must be filled with some impermeable materials, such as thermostable mineral wool. This approach was used, for example, in the production process of the catalytic incineration of organic admixtures in waste gases[22]; however, it significantly complicates the GFC loading and replacement procedures.

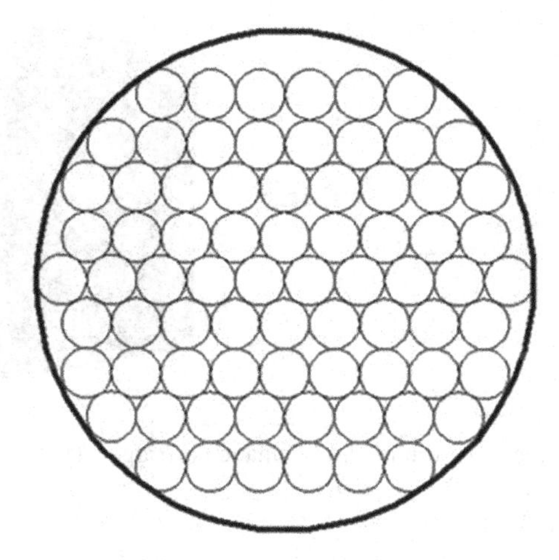

**FIGURE 3.8** Catalyst bed in the round reactor, formed from spiral GFC cartridges (top view).

### 3.3.2. Prismatic Cartridges

To overcome the problems just discussed, it is possible to use cartridges of prismatic shapes[23–26] (e.g., with a square, rectangle, triangle and trapezium sequence; Figure 3.9).

Prismatic GFC cartridge consists of the catalyst clothes, plain and corrugated structuring metal meshes (Figure 3.10). The plain mesh is placed between the layers of GFC fabric to create a plain and mechanically stable catalytic elements. Corrugated mesh is laid between such elements to make the passages for the transportation of the reaction media. Outside all these elements are fixed by the outer body, also made of metal mesh.

The corrugated mesh, except forming the passages for the reaction mixture, also provides the mechanical stability of the system in general, especially its resistance to vibration and mechanical shocks.

At the same time, the corrugated meshes may cause an excessive pressure drop of the cartridges. They may also occupy the greater part of the used metal mesh, thus causing increased metal consumption, and therefore excessive cost of the cartridges.

It is possible to solve this problem by the application of corrugation-free cartridges,[26] where the GFC is surrounded by the two external metal meshes, thus forming the plain catalytic elements. These elements are located in parallel with each other, with their edges fixed at the side walls of the external cartridge body to provide the overall mechanical strength and rigidity of the whole structure (Figure 3.11).

In general, all prismatic cartridges, as well as spiral ones, provide the efficient contact between the reaction media and catalyst. They are characterized with a low pressure drop and high mechanical stability and resistance to mechanical and thermal shocks. In addition, they may be applied in dusty media, in particular, in the presence of soot particulates.

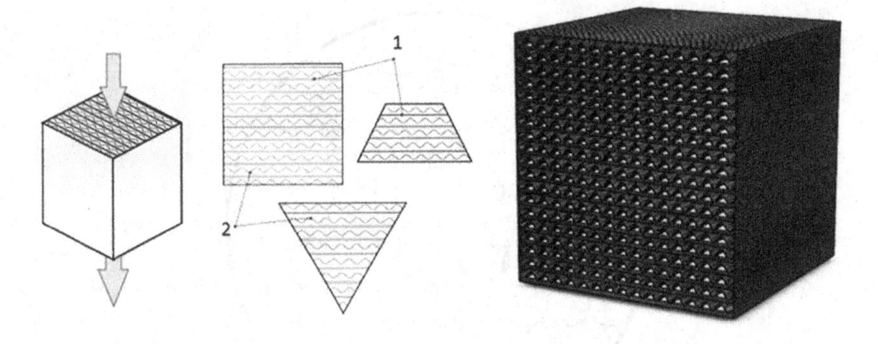


**FIGURE 3.9** Schematic structure of the prismatic cartridges: (1) GFC and (2) volumetric structuring elements.

**FIGURE 3.10** Internal structure (on the left) and external view of the prismatic cartridge with the corrugated structuring elements.

**FIGURE 3.11** Schematic structure and appearance of corrugation-free cartridge.

Figure 3.12 shows the examples of formation of different modular beds from the prismatic cartridges in the large-scale reactors, including those with non-cylindrical shape. It is seen that the application of the prismatic cartridges provides more dense bed packing, completely excluding the void between cartridges and minimizing the voids on the connection between the cartridges and reactor walls. The most impressive effect was achieved in the loading of reactors with a square of rectangle shape (Figure 3.12c), where the voids in the wall vicinity could be completely removed.

Cartridges with a square sequence are the simplest to produce. We selected cubic blocks of 160 × 160 × 160 mm as a basic size and shape. They are convenient both in manufacturing and handling. They are also compatible with the standard catalytic monolith of 6 × 6 × 6 inches, widely distributed in the USA and worldwide.

These cartridges may be used for the formation of modular GFC beds of practically any size and shape. Thereby, they are most suitable for the arrangement of beds with axial movement of gliding flow. However, in some cases, such as in the case of strong limitations on the maximum pressure drop in a reactor, they may be arranged into the modular radial bed. In this bed, cartridges can be assembled into radial bed, though the reaction mixture fluid inside each cartridge will be an axial-gliding type.[16,27,28] Unlike the multilayer radial bed in Figure 3.6, this modular bed is applicable in large-scale reactors and in the treatment of dusty fluids, where the gliding

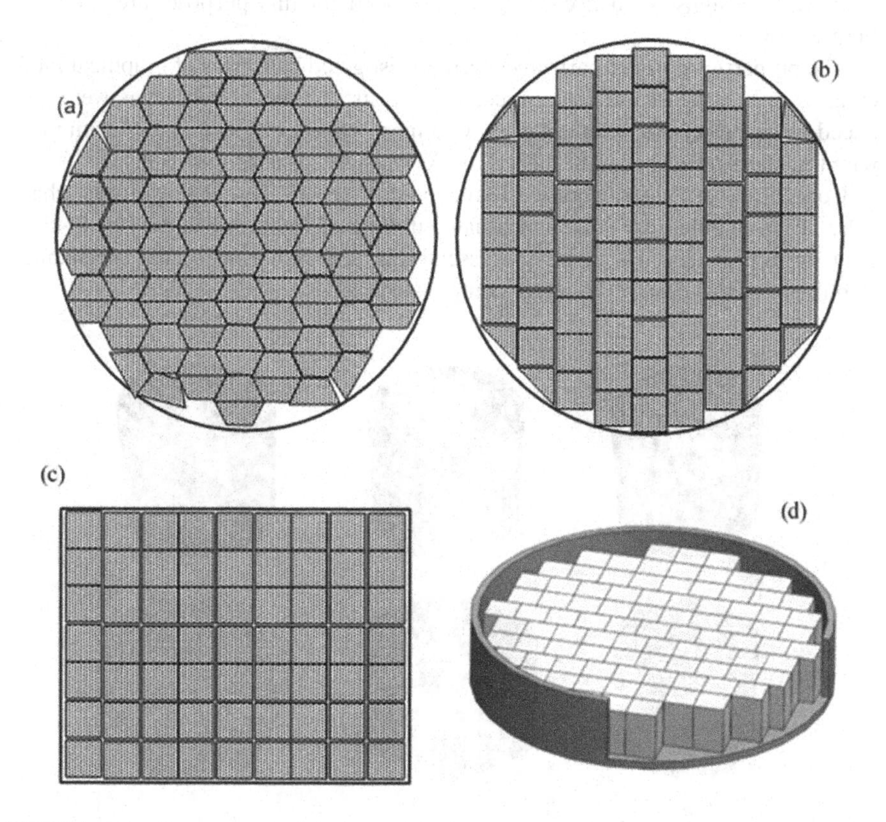


**FIGURE 3.12** Structure of catalytic modular beds formed from prismatic GFC cartridges.

flow regime is advantageous. It is also characterized with a quite low pressure drop. The example of practical application of such beds is described in Chapter 5.1.2.

### 3.3.3. Reinforced Cartridges

GFC-based catalytic cartridges may be applied in the processes of the combustion of dispersed solid fuels in the fluidized bed of heat transfer particles[29] (see Chapter 5.2.1). This process is specific as the heat transfer agent, such as sand, has a high susceptibility to abrasion and mechanical attrition of the fixed catalyst beds. Therefore, this process imposes rather high requirements to the catalysts regarding their resistance to attrition. The second problem is a possibility of clogging the channels in the catalyst block with the particles of both the fuel and heat-transfer agent, making necessary the development of well-structured cartridges with wide channels.

To solve these problems, we have proposed reinforced GFC cartridges, where the GFC fabric is protected from the attrition by the external layers of the protecting metal wire mesh.[26] The size of the cell in the mesh should be less than typical size of heat-transfer particles to exclude the mechanical contact between these particles and catalyst cloth, thus protecting the GFC from abrasion. Such cell size should be sufficient to provide the access of gaseous reactant to the GFC surface.

Possible designs of such cartridges, developed for this purpose, are shown in Figure 3.13.

The properties of these cartridges were investigated by means of graphical modeling as well as by experimental mechanical tests; the model cartridges were produced using protective stainless-steel wire mesh with a cell size of 2 × 2 mm and wire thickness of 0.32 mm. IC-12-S111 GFC was used as a sample catalyst.

Concentric cartridges (Figure 3.13a) consist of several nested ring-shaped channels. These channels are wide and situated uniformly to minimize the risk of clogging by particulates. However, the assembling and provision of the mechanical stability of such a design is a rather complicated task.

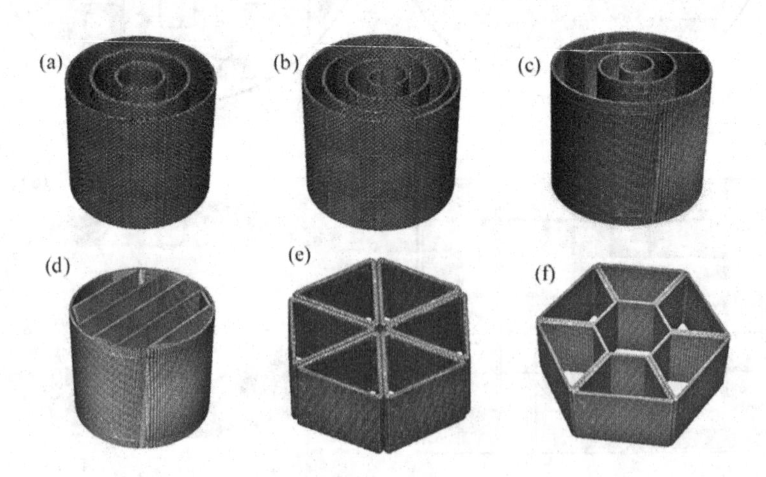


**FIGURE 3.13** Reinforced GFC cartridges.

**FIGURE 3.14** Multicell modular-reinforced GFC cartridge.

Spiral reinforced cartridges (Figures 3.13b and 3.13c) are easier to produce. Graphical modeling of a cartridge with an Archimedes spiral (Figure 3.13b) showed the presence of thin dead-end zones with a high clogging risk, though these zones can be excluded by the modification of spiral geometry (Figure 3.13c). The common problem of the spiral cartridge is that it is complicated to provide the spatial uniformity and stability of channel width, especially taking into account the probable deformation of protecting meshes under exposure to high temperatures in the process conditions.

To some extent, this problem is solved in a cartridge with the parallel positioning of the reinforced GFC planes (Figure 3.13d). These cartridges are more stable due to the mounting of these planes to the round outer wall, though such stabilization is efficient only for the small-scale blocks, and the problem of plane deformation remains in case of larger size.

The efficient scaling up can be provided by assembling the modular catalyst beds from simple and small elements. For example, it is possible to make the beds of elemental reinforced GFC prisms of triangle (Figure 3.13e) or trapezoid (Figure 3.13f) shape. Figure 3.14 shows the design of the multicell cartridge made of the triangle-reinforced GFC prisms.

## 3.4. LEMNISCATE GFCs

The structuring of GFC microfibers in the form of woven or knitted cloths provides good strength and flexibility of the catalytic material. However, in the course of reaction the transportation of reactants from moving fluid to fiber surface may be deteriorated in the interlacing zone by mutually overlapping catalyst threads. These interlacing zones may involve a significant amount (up to 50% and more) of the cloth surface area; therefore, this may cause a decrease of the external mass transfer intensity. Moreover, during the supporting of the active component on the microfiber fabric surface by the means of impregnation method, the major part of the precursor solution may accumulate in the thread interlacing area due to the influence of surface tension forces.[30] All in all, it leads to a situation when the major amount of the active component is in the GFC area, which is less available for reactants. It may cause a significant decrease of apparent GFC activity due to inefficient use of the catalytic potential of the active component. Therefore, the development of new geometrical forms of microfibrous materials with a minimized or completely excluded area of fibers and threads interlacement is an important task.

An additional challenge is the decrease of the metal consumption for structuring of the GFC packings. The structuring metal meshes, both plain and volumetric, take up to 90% of the total cartridge weight. This is a serious drawback, as it complicates the manufacturing of GFC cartridges and increases their cost, especially in cases of the application of expensive metals and alloys with the improved corrosion resistance and thermal stability, such as stainless steel or special thermostable alloys. In total, the cost of structuring elements may exceed 50% of the total cartridge cost, being higher even than that for the active noble metals.

These problems can be solved by the application of the novel GFC structure.[31,32] It consists of threads twisted from glass microfibers in a shape of either closed (Figure 3.15a) or open (Figure 3.15b) loop or mono-focus lemniscate, as well as two-focus lemniscate (Figure 3.15c). Each element can be attached to one (Figure 3.15d) or two (Figure 3.15e) plain supporting cloths.

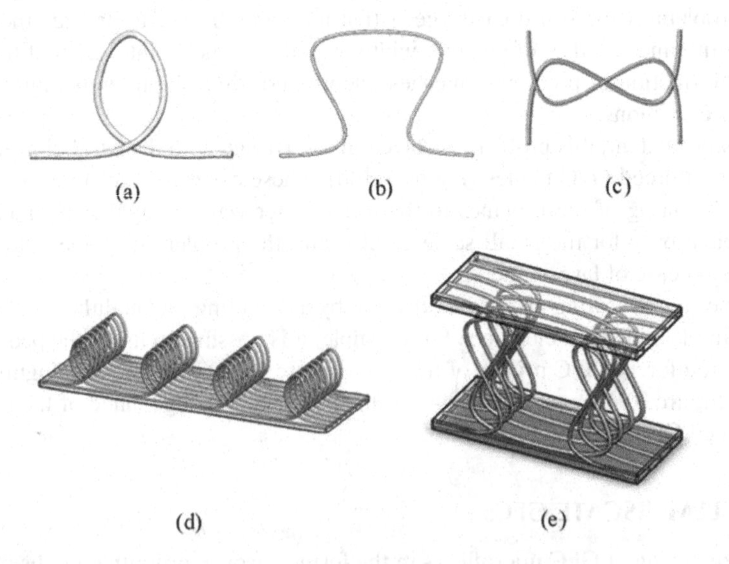


**FIGURE 3.15** External view of GFC elements in the form of mono-focus (a, b) and two-focus lemniscates (c) and their combination with the supporting cloths (d, e).

*Source:* Reprinted with permission from Lopatin et al. 2017.[32]

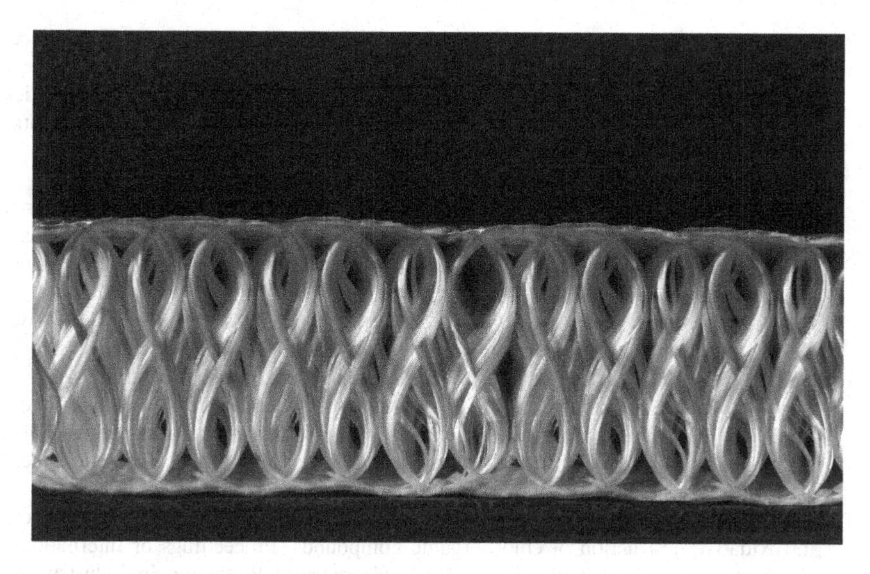

**FIGURE 3.16** Lemniscate GFC.

*Source:* Photo by Eugenia Bobatkova.

An external view of lemniscate GFC element is shown in Figure 3.16. These elements can be assembled into cartridges of either round or prismatic shape.[31] Such cartridges are mechanically stable, flexible, resilient, and highly permeable for the reacting fluid. They have a high structural uniformity, resulting in a low pressure drop. Notably, they are structurally self-sufficient, and all these benefits can be achieved without use or with a minimum use of the additional structuring elements. It helps to simplify the manufacturing of GFC cartridges and decrease their weight and cost.

## REFERENCES


1. Zagoruiko A.N., Shinkarev V.V., Simonova L.G. 2007a. Vanadia/glass-fiber catalyst for hydrogen sulfide selective oxidation to sulfur by oxygen. Proceedings of 3rd International Conference "Catalysis: Fundamentals and Application" (Novosibirsk, Russia, July 4–8): 2, 580.
2. Cheng X., Yang H., Tatarchuk B.J. 2016. Microfibrous entrapped hybrid iron-based catalysts for Fischer—Tropsch synthesis. *Catal Today*. 273: 62–71.
3. Lopatin S.A., Zagoruiko A.N. 2014. Pressure drop of structured cartridges with fiberglass catalysts. *Chem Eng J.* 238: 31–6.
4. Veniaminov S.A., Zagoruiko A.N., Simonova L.G. et al. 2005. Investigation of carbon monooxide oxidation over Pt, Au and Pt-Au containing catalysts with flass-fibre support. Proceedings of 7th European Congress on Catalysis (Sofia, Bulgaria, August 28–September 1): 9–25.
5. Zagoruiko A.N., Veniaminov S.A., Veniaminova I.N. et al. 2006. Kinetic instabilities and intra-thread diffusion limitations in CO oxidation reaction at Pt/fiber-glass

catalysts. Proceedings of International Conference "Chemreactor-17" (Athens, Greece, May 2006): 604–7.
6. Zagoruiko A.N., Veniaminov S.A., Veniaminova I.N. et al. 2007. Kinetic instabilities and intra-thread diffusion limitations in CO oxidation reaction at Pt/fiber-glass catalysts. *Chem Eng J.* 134: 111–16.
7. Nakamura H. 1974. British Patent No.1372806. Exhaust gas purifying apparatus.
8. Zagoruiko A.N., Glotov V.D., Lopatin S.A. et al. 2016. Investigation of the internal structure, fluid flow dynamics and mass transfer in the multi-layered packing of glass-fiber catalyst in the pilot reactor for sulfur dioxide oxidation. *Science Bulletin of the Novosibirsk State Technical University.* 3(64): 161–77.
9. Barelko V.V., Prudnikov A.A., Bykov L.A. et al. 2001. Russian Patent No.2171430. Device for thermocatalytic purification of vent exhausts from painting chambers.
10. Nakamura H. 1975. US Patent No.3897366. Automotive exhaust gas catalyst.
11. Zagoruiko A.N., Bal'zhinimaev B.S., Arendarskii D.A. et al. 2004. Russian Patent No.2231653. Device for purification of exhaust gases from internal combustion engines.
12. Zagoruiko A.N., Bal'zhinimaev B.S., Arendarskii D.A. 2003. Russian Patent No.2200622. Method for performance of heterogeneous catalytic reactions.
13. Zagoruiko A.N., Bal'zhinimaev B.S., Beskopylnii et al. 2007. Pilot tests of the process for oxidative destruction of chloroorganic compounds. Proceedings of International Conference "Catalytic Technologies for Environmental Protection in Industry and Transport" (Saint-Petersburg, December 11–14): 198–9.
14. Arendarskiy D.A., Zagoruyko A.N., Bal'zhinimaev B.S. 2005. Glass-fibre catalysts to clear diesel engine exhausts. *Chemistry for Sustainable Development.* 13(6): 731–5.
15. Zagoruiko A.N., Bal'zhinimaev B.S. 2011. Catalytic processes on the base of glass-fiber catalysts. *Chemical Industry Today.* 2: 5–11.
16. Zagoruiko A.N., Lopatin S.A., Klenov O.P. 2013. Russian Patent for Utility Model No.124888. Reactor for performance of heterogeneous catalytic process.
17. Zagoruiko A.N., Arendarskii D.A., Bal'zhinimaev B.S. 2007. Russian Patent for Utility Model No.66975. Catalytic system for performance of heterogeneous reactions.
18. Zagoruiko A.N., Lopatin S.A., Zazhigalov S.V. et al. 2018. Russian Patent for Utility Model No.176547. Catalytic block for performance of heterogeneous reactions.
19. Zagoruiko A.N., Lopatin S.A., Zazhigalov S.V. et al. 2018. Russian Patent for Utility Model No.177155. Block for performance of catalytic heterogeneous reactions.
20. Zagoruiko A.N., Lopatin S.A., Zazhigalov S.V. et al. 2018. Russian Patent for Utility Model No.177270. Block for performance of catalytic heterogeneous reactions.
21. Bal'zhinimaev B.S., Zagoruiko A.N., Arendarskii D.A. 2007. Russian Patent for Utility Model No.66974. Catalytic system for performance of thermally intensive reactions.
22. Zagoruiko A.N., Lopatin S.A., Bal'zhinimaev B.S. et al. 2010. The process for catalytic incineration of waste gas on IC-12-S102 platinum glass fiber catalyst. *Catalysis in Industry.* 2: 113–17.
23. Zagoruiko A.N., Lopatin S.A., Zazhigalov S.V. et al. 2017. Russian Patent for Utility Model No.174588. Catalytic block for performance of heterogeneous reactions.
24. Zagoruiko A.N., Lopatin S.A., Bal'zhinimaev B.S. et al. 2011. Russian Patent for Utility Model No.101653. Catalytic block and catalytic system for incineration of hazardous organic admixtures in waste gases.
25. Zagoruiko A.N., Lopatin S.A., Bal'zhinimaev B.S. 2011. Russian Patent for Utility Model No.101652. Catalytic block and catalytic system for performance of heterogeneous catalytic reactions.
26. Zagoruiko A.N., Lopatin S.A. 2014. Russian Patent for Utility Model No.145037. Catalytic cartridge for performance of heterogeneous catalytic reactions.
27. Zagoruiko A.N., Lopatin S.A., Klenov O.P. 2013. Russian Patent for Utility Model No.125094. Catalytic system for performance of heterogeneous reactions.

28. Serbinenko V.V., Zagoruiko A.N., Lopatin S.A. et al. 2013. Russian Patent for Utility Model No.124925. Catalytic system for purification of diesel exhausts.
29. Lopatin S., Chub O., Yazykov N. et al. 2014. Structured cartridges with reinforced fiber-glass catalyst for fuel combustion in the fluidized beds of the inert heat-transfer particles. Proceedings of the XXI International Conference on Chemical Reactors "CHEMREACTOR-21" (Delft, The Netherlands, September 22–25): 272–3.
30. Chub O.V. 2009. Investigation of mass transfer processes in glass-fiber catalytic systems. PhD Thesis. Novosibirsk: Boreskov Institute of Catalysis.
31. Zagoruiko A.N., Lopatin S.A., Zazhigalov S.V. et al. 2016. Russian Patent for Utility Model No.166263. Catalytic cartridge for performance of heterogeneous catalytic reactions.
32. Lopatin S.A., Mikenin P.E., Pisarev D.A. et al. 2017. A microfiber catalyst with lemniscate structural elements. *Catalysis in Industry*. 9(1): 39–47.

# 4 Experimental Investigation of Pressure Drop and Mass Transfer in GFC Packing

In general, the traditional design of the existing catalytic reactors is not applicable, or at least not optimal, for the application of GFCs due to their unusual shape and mechanical properties. The development of a new design, optimized for GFC use, should be based on the scientific engineering approach, but this basis was not prepared prior to our research.

Earlier,[1,2] mass transfer processes in GFC beds were studied experimentally, using the model reaction of CO oxidation at GFC; these experiments resulted in the formulation of criteria equations for the calculation of the mass transfer coefficients. However, the studies involved only multilayer GFC packing, with the propagative flow of reaction fluid through these layers. In addition, the issue of pressure drop in such beds was not discussed. Practically no information on the mass transfer and pressure drop of GFC packing with a gliding flow is available in the scientific and technical literature at that moment.

## 4.1. THE SCOPE AND PROPERTIES OF THE RESEARCH OBJECTS

The comparative studies of the mass transfer and hydraulic resistance involved various types of GFC structures, which were described in Chapter 3:

- Cartridges with the corrugated metal mesh structuring elements
- Cartridges with the plain metal mesh structuring elements
- GFC beds with the lemniscate structure
- Transverse multilayered GFC bed

The described commercial Pt/GFC IC-12-S111 with an average platinum content of 0.07% was used in all these catalytic systems. Stainless steel mesh with a cell size 1 × 1 mm made of 0.32-mm wire was utilized for the construction of both corrugated and plain structuring elements.

### 4.4.1. Experimental GFC Cartridges With the Corrugated Metal Mesh Structuring Elements

Experimental GFC cartridges with the corrugated metal mesh structuring elements represented the cubes 44 mm (for kinetic and mass transfer experiments) or 160 mm (for pressure drop studies) on edge. Four cartridges with the different channel height (3, 5, 7, and 9 mm) were produced for the experiments (Figure 4.1).

The internal structure of such cartridges consists of alternating layers: "plain mesh—catalytic fabric—corrugated mesh—catalyst fabric" (see Figure 4.2).

Assembly of the necessary number of such layers, required for the formation of the cartridge with a given height, is placed inside the casing made of plain mesh, which covers the upper, bottom, and side walls of the cartridge, leaving the front and back sides uncovered to provide unimpeded access of the reaction fluid into the cartridge.

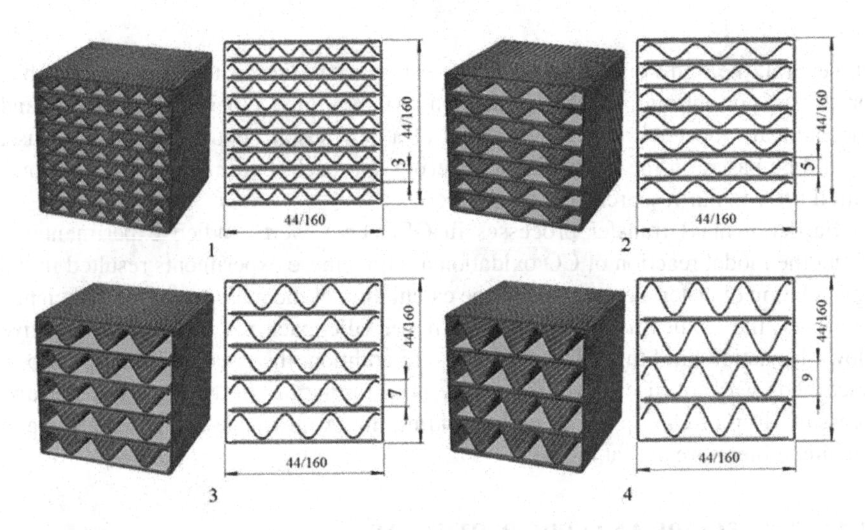


**FIGURE 4.1** GFC cartridges with the corrugated metal mesh structuring elements and different channel heights: (1) 3 mm, (2) 5 mm, (3) 7 mm, (4) 9 mm.

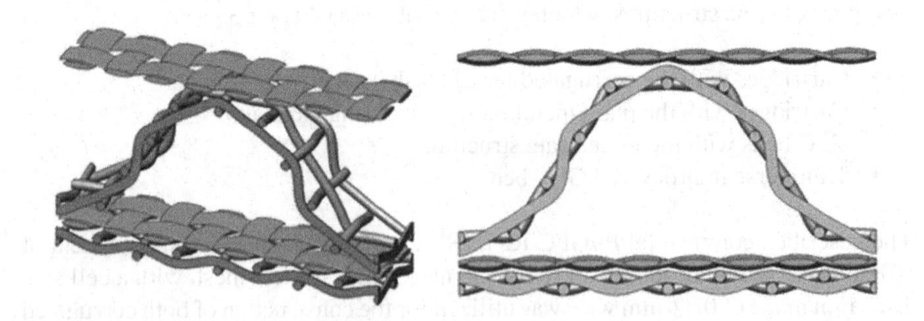

**FIGURE 4.2** Elementary cell of GFC cartridge with the corrugated metal mesh structuring elements.

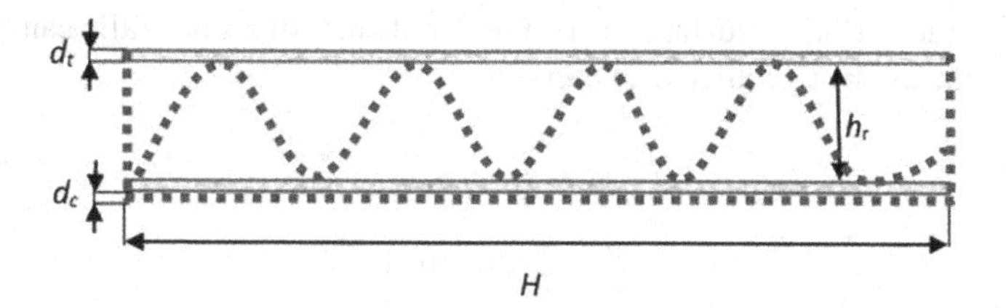


**FIGURE 4.3** Unit layer of the GFC cartridge.

The specific area of the cartridge external surface is equal to the ratio between the surface area of the washable cartridge elements and its volume. For simplicity, we may consider one cartridge layer with a width $H$, consisting of one corrugated element with the corrugation height $h_g$, $n_c$ of plain meshes with thickness $d_c$, $n_m$, GFC fabric with the thickness $d_m$ and two side walls also made of plain mesh with the thickness $d_c$, and total height $h = h_g + n_c\, d_c + n_m\, d_m$.

An example of such a layer for two GFC fabrics ($n_t$ = 2) and one plain mesh ($n_c$ = 1) is shown in Figure 4.3.

In this case the washable surface includes the external surface of wire in a corrugated mesh $S_g$; two surfaces of GFC fabric, oriented towards corrugation $S_\kappa$; and internal surfaces of the side walls $S_w$.

The specific surface area of wire in corrugation can be calculated from the wire diameter $d$ (m); mesh density $M$ (kg/m$^2$), which is specified by the mesh manufacturer; and the bulk density of metal used for mesh manufacturing, $\rho$ (kg/m$^3$). Obviously, the metal volume ($v_m$) in 1 m$^2$ of mesh is equal to $v_m = M/\rho$ (m$^3$/m$^2$); and for the wire with the constant round sequence of the bulk metal volume in 1 m$^2$ of mesh is equal to $\pi d^2 l/4$, where $l$ is a total wire length in 1 m$^2$ of mesh (m/m$^2$).

From the equation $\frac{M}{\rho} = \frac{\pi d^2 l}{4}$ it is possible to derive the value of l: $l = \frac{4M}{\pi d^2 \rho}$ (m/m$^2$). Surface area of wire in 1 m$^2$ of mesh will be:

$$s = \pi d l = \frac{4M\pi d}{\pi d^2 \rho} = \frac{4M}{d\rho}, \left(\mathrm{m}^2/\mathrm{m}^2\right) \tag{4.1}$$

Taking into account that the linear length of the corrugated mesh is more than the one for plain mesh, it is necessary to multiply this value by the mesh lengthening ratio $\varphi$ (empirical coefficient, depending on the real geometry of the given corrugation; for typical corrugations it is in the range of 1.2–1.4:

$$s = \frac{4\varphi M}{d\rho}, \left(\mathrm{m}^2/\mathrm{m}^2\right) \tag{4.2}$$

For the layer width $H$ and its depth $L$ the area of the horizontal cartridge sequence is equal to $HL$. The surface area of corrugation in this sequence is equal to:

$$S_g = \frac{4\varphi MHL}{d\rho}, \left(\mathrm{m}^2\right) \tag{4.3}$$

The area of the washable surface of the GFC and cartridge side walls can be calculated from the evident geometric expressions:

$$S_K = HL, (\mathrm{m}^2) \tag{4.4}$$

$$S_w = h_T L, \left(\mathrm{m}^2\right) \tag{4.5}$$

The full volume of the cartridge layer is equal to:

$$V = HL\left(h_g + n_c d_c + n_T d_T\right), \left(\mathrm{m}^3\right) \tag{4.6}$$

Respectively, the specific surface area is equal to:

$$S_{SP} = \frac{S_g + 2S_\kappa + 2S_w}{V} = \frac{4\varphi MHL/\rho d + 2HL + 2h_g L}{HL\left(h_g + n_c d_c + n_T d_T\right)} = \frac{4\varphi M/\rho d + 2HL + 2h_g/H}{h_g + n_c d_c + n_T d_T}, \left(\mathrm{m}^{-1}\right) \tag{4.7}$$

To calculate the void fraction of the cartridge it is necessary to find the specific volume of all structural bed elements: corrugation $v_g$, catalyst $v_\kappa$, plain mesh $v_c$, and side walls $v_w$. The mass of metal in the corrugation equals to $HL\varphi M$; therefore, the total wire volume in the corrugation is:

$$v_g = \frac{HL\varphi M}{\rho}, \left(\mathrm{m}^3\right) \tag{4.8}$$

The volumes of catalyst, plain mesh, and side walls are defined from geometrical considerations:

$$v_\kappa = n_T HL d_T, (\mathrm{m}^3) \tag{4.9}$$

$$v_C = HL d_C, \left(\mathrm{m}^3\right) \tag{4.10}$$

$$v_w = \left(h_g + n_c d_c + n_T d_T\right) d_c L, \left(\mathrm{m}^3\right) \tag{4.11}$$

Bed void fraction is defined as $\varepsilon = 1 - \frac{v_g + v_\kappa + v_C + 2v_w}{V}$, correspondingly:

$$\varepsilon = 1 - \frac{HL\varphi M/\rho + n_T HL d_T + n_C HL d_C + 2\left(h_g + n_c d_c + n_T d_T\right) d_c L}{HL\left(h_g + n_c d_c + n_T d_T\right)} = 1 - \frac{\varphi M/\rho + n_c d_c + n_T d_T}{h_g + n_c d_c + n_T d_T} - \frac{2d_c}{H} \tag{4.12}$$

These equations can be used for the calculation of $S_{SP}$ and $\varepsilon$ both for cartridges with the corrugated structuring meshes and for the volumetric chain link nets. In the latter case, coefficient $\varphi$ should be taken equal $\varphi = 1$, as long as the density of 1 $m^2$ of net $M$ is given in the reference literature already with the account of real wire geometry in the net. These equations can be also applied to calculate the cartridges with the plain metal mesh structuring element, setting $\varphi = 0$ for this case.

For the cylinder block, assuming that cartridge radius is much more than the layer height (i.e., that it is possible to neglect the layer curvature), we can use the same approach, excluding side walls from the consideration ($S_w = 0$, $v_w = 0$); therefore:

$$S_{SP} = \frac{4\varphi M / \rho d + 2}{h_g + n_c d_c + n_T d_T}, \left(m^{-1}\right) \tag{4.13}$$

$$\varepsilon = 1 - \frac{\varphi M / \rho + n_c d_c + n_T d_T}{h_g + n_c d_c + n_T d_T} \tag{4.14}$$

The equivalent hydrodynamic diameter of a channel in a cartridge $d_{eq}$ was calculated using the equation:

$$d_{eq} = \frac{4\pi\varepsilon}{S_{SP}}, (m) \tag{4.15}$$

Notably, in the case of cartridges with the corrugated structuring elements, their total washable surface area is not equivalent to the specific external surface area of the GFC as long as the significant part of the washable surface belongs to the corrugated mesh. Thus, the area of the washable GFC surface $S^*_{SP}$ was calculated using the modified equation (4.13):

$$S^*_{SP} = \frac{2h_g / H}{h_g + n_c d_c + n_T d_T}, \left(m^{-1}\right) \tag{4.16}$$

The equivalent hydrodynamic diameter of the cartridge channel in the mass transfer studies $d^*_{eq}$ was calculated from empirical equation[3]:

$$d^*_{eq} = 1.265 \times \left(\frac{H^3 h_g^3}{H + h_g}\right)^{1/5}, (m) \tag{4.17}$$

The values $S_{SP}$ and $d_{eq}$ were later used in the calculation of Reynolds criterion, characterizing the fluid turbulization degree, while the calculation of Sherwood criterion was based on the corresponding values of $S^*_{SP}$ and $d^*_{eq}$.

Technical and geometric characteristics of the cartridges with the corrugated metal mesh structuring elements are given in Table 4.1.

**TABLE 4.1**
**Properties of GFC Cartridges with Corrugated and Flat Mesh Structuring Elements**

| Parameter | Passage height (*h*) | | | |
|---|---|---|---|---|
| | 3 mm | 5 mm | 7 mm | 9 mm |
| Number of GFC layers | 18 | 14 | 10 | 8 |
| Number of mesh layers | 10 | 8 | 6 | 5 |
| Catalyst mass in the cartridge, $m_{cat}$, g | 8.7 | 7.14 | 4.91 | 3.77 |
| Active component mass in the cartridge, $m_{ac}$, g | 0.00609 | 0.005 | 0.00344 | 0.00264 |
| Porosity, $\varepsilon$ | 0.631 | 0.710 | 0.782 | 0.817 |
| Specific cartridge external surface area, $S_{sp}$, m$^{-1}$ | 885 | 666 | 531 | 472 |
| Specific external surface area of GFC, $S^*_{sp}$, m$^{-1}$ | 338 | 263 | 188 | 150 |
| Equivalent pass diameter for $Sh$ number calculations, $d_{eq}$, m | 0.01139 | 0.01397 | 0.0179 | 0.02081 |
| Equivalent pass diameter for *Re* number calculations, $d^*_{eq}$, m | 0.00285 | 0.00455 | 0.00589 | 0.00692 |

*Source*: Reprinted with permission from Zagoruiko et al. 2017.[4]

### 4.1.2. GFC Beds With the Lemniscate Structures

Experimental cartridges of the lemniscate structure were manufactured from the GFC layers with loop-type threads (Figure 4.4).

Each cartridge consisted of few such layers, assembled inside the metal mesh external housing. In total, three different cartridges were made of lemniscate GFC with the passage heights of 8, 10, and 15 mm (see Figure 4.5).

The calculation of the geometric parameters of these cartridges was performed by the graphical modeling method using the SolidWorks 2015 software. One volumetric segment was considered in these calculations (Figure 4.6). Calculated parameters are presented in Table 4.2.

### 4.1.3. GFC Cartridges With the Flat Mesh Structuring Elements

In these cartridges the GFC layer was made in the form of a "sandwich" with the two fabrics of GFC and plain metal mesh between them (Figure 4.7); the design was fixed with brackets to provide its firmness.

The layers were assembled into the cartridge in parallel to each other and fixed on the side walls. Two cartridges with the interlayer distance *h* equal to 3 and 5 mm were used in the experiments (Figure 4.8).

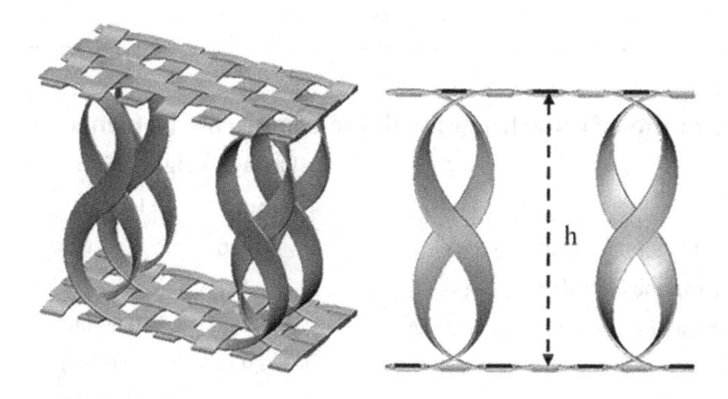


**FIGURE 4.4** Structure of lemniscate GFC.

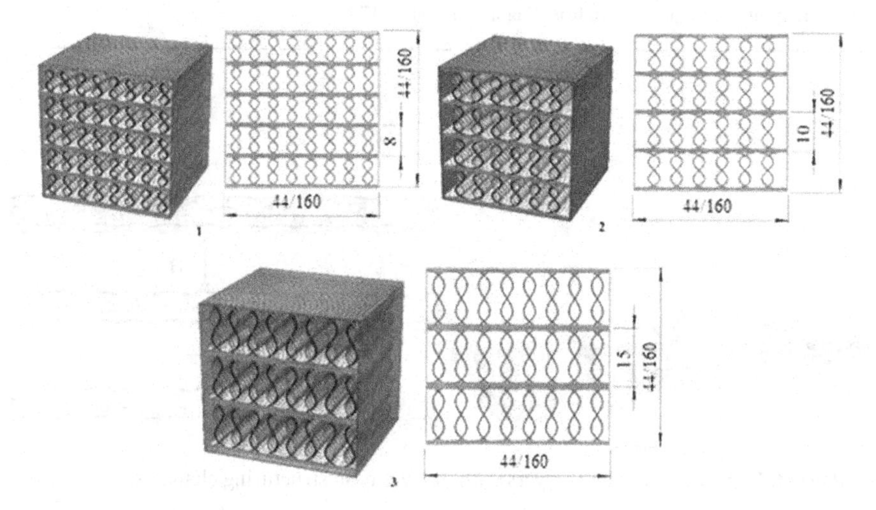


**FIGURE 4.5** Cartridges on the base lemniscate GFC with the different layer heights: (1) 8 mm, (2) 10 mm, (3) 15 mm.

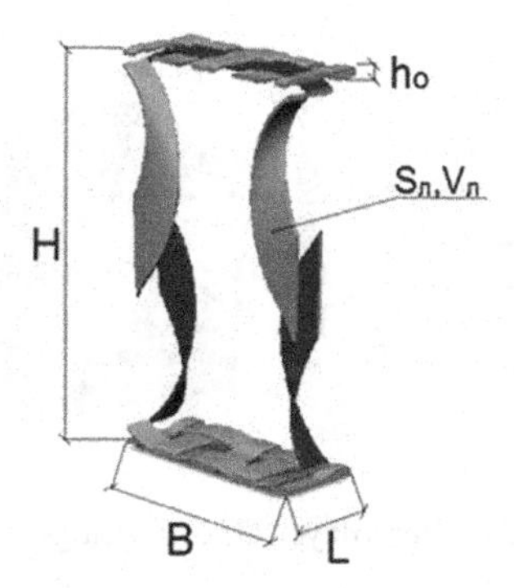


**FIGURE 4.6** Elementary volume element of lemniscate GFC. $H$ = height of lemniscate layer, $B$ = width, $L$ = depth, $h_0$ = basement thickness.

**TABLE 4.2**
**Properties of the GFC Cartridges with the Lemniscate Elements**

| Parameters | Passage height ($H$) | | |
|---|---|---|---|
| | **8 mm** | **10 mm** | **15 mm** |
| Number of GFC lemniscate layers | 6 | 4 | 3 |
| Catalyst mass in the cartridge, $m_{cat}$, g | 9.7 | 7.2 | 6.27 |
| Active component mass in the cartridge, $m_{ac}$, g | 0.00728 | 0.00540 | 0.00470 |
| Porosity, $\varepsilon$ | 0.933 | 0.920 | 0.838 |
| Specific external surface area of GFC, $S_{sp}$, m$^{-1}$ | 370 | 430 | 301 |
| Equivalent pass diameter, $d_{eq}$, m | 0.01010 | 0.00850 | 0.01110 |

*Source*: Reprinted with permission from Zagoruiko et al. 2017.[4]

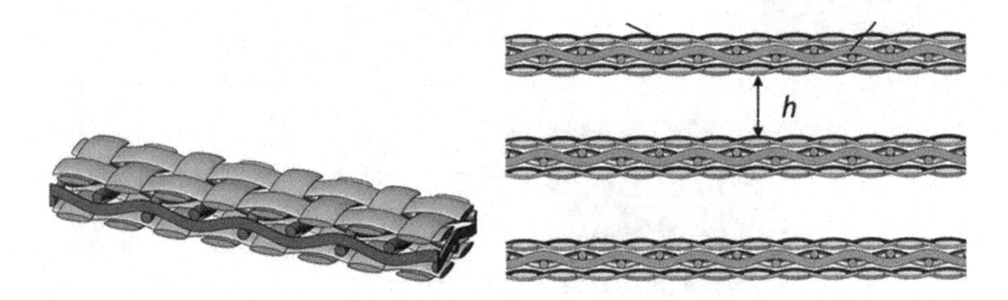


**FIGURE 4.7** Structure of GFC layers with the flat mesh structuring elements.

*Source*: Reprinted with permission from Zagoruiko et al. 2017.[4]

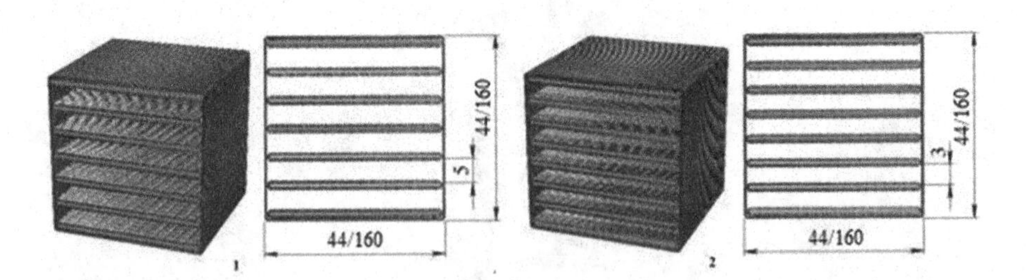


**FIGURE 4.8** Appearance of the multilayer GFC cartridge with the interlayer distance of 3 mm (1) and 5 mm (2).

**TABLE 4.3**
**Properties of the GFC Cartridges with the Flat Mesh Structuring Elements**

| Parameter | Passage height (*h*) | |
|---|---|---|
| | 5 mm | 3 mm |
| Number of GFC layers | 12 | 18 |
| Number of mesh layers | 5 | 8 |
| Catalyst mass in the cartridge, $m_{cat}$, g | 6.13 | 8.8 |
| Active component mass in the cartridge, $m_{ac}$, g | 0.00429 | 0.00616 |
| Porosity, $\varepsilon$ | 0.856 | 0.721 |
| Specific external surface area of GFC, $S_{sp}$, $m^{-1}$ | 225 | 338 |
| Equivalent channel diameter, $d_{eq}$, m | 0.015118 | 0.00854 |

*Source*: Reprinted with permission from Zagoruiko et al. 2017.[4]

The properties of these cartridges are given in Table 4.3.

### 4.1.4. Reference Catalysts

The additional reference catalysts were included into the experimental studies to provide the direct and accurate comparison of the proposed GFC structures.

First, it was decided to compare the efficiency of gliding and propagative fluid arrangements in GFC structures. A system of 10 parallel layers of IC-12-S111 GFC was produced for this purpose. These layers were placed into the metal mesh body with a size 44 × 44 × 44 mm and plain supporting metal mesh, attached to the side walls of this body (Figure 4.9). Properties of this packing are described in Table 4.4.

The spiral block was used in the experiments in addition to the prismatic cartridges to study the influence of the cartridge external shape on its pressure drop. These cartridges were described in Chapter 3.3.1, their view is shown in Figure 4.10.

To compare the mass transfer efficiency in the different structures, the conventional catalysts were also included into the experimental program, namely: granular catalyst, honeycomb monolith, and wire-mesh blocks.

The appearance of monoliths is given in Figure 4.11. Two different monolith blocks with alumina support having square channels were used, with a different channel size. Platinum was selected as an active component, and to reproduce the similar intrinsic activity the procedure of Pt supporting at $Al_2O_3$ block was the same as for the production of GFC IC-12-S111.

A commercial Pt-$Al_2O_3$ catalyst (trademark AP-64, produced by Novokuibyshevsk Catalyst Plant) was selected as the sample granular catalyst for the comparative studies (Figure 4.12). Catalyst granules had a shape of cylinders with the diameter of ~ 3 mm and height of ~ 1 cm. In the experiments, they were packed into the cubic container made of metal mesh.

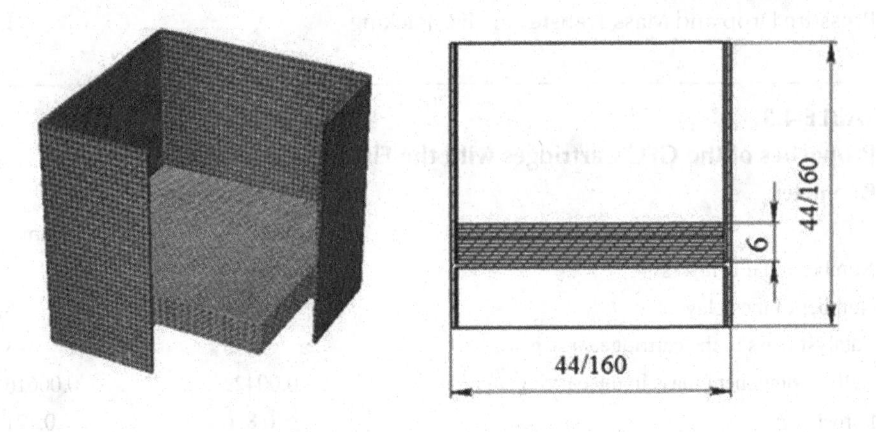


**FIGURE 4.9** Multilayered GFC stack for the propagative fluid arrangement.

**TABLE 4.4**
**Properties of the Multilayered GFC Packing**

| Parameter | Value |
|---|---|
| Number of GFC fabric layers | 10 |
| Catalyst mass in the cartridge, $m_{cat}$, g | 4.7 |
| Active component mass in the cartridge, $m_{ac}$, g | 0.00329 |
| Porosity, $\varepsilon$ | 0.707 |
| Specific area of washable surface of fibers, $S_{SP}$, m$^{-1}$ | 1.95*10$^5$ |
| Equivalent pass diameter, $d_{eq}$, m | 0.00040 |

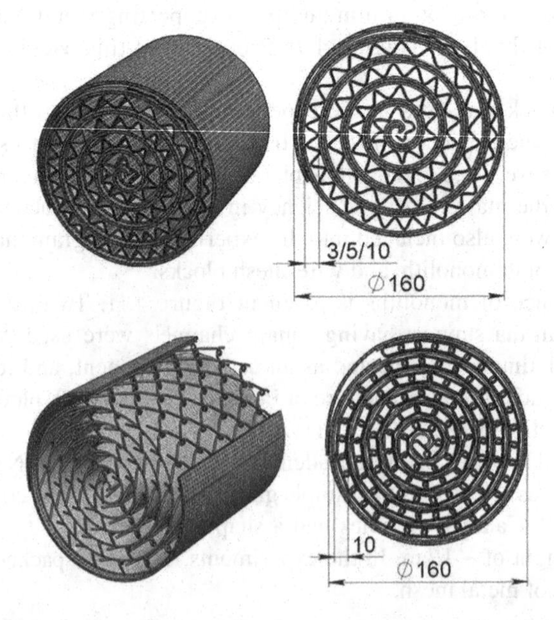


**FIGURE 4.10** Spiral GFC cartridges with the corrugated (above) and chain-link (below) structuring elements.

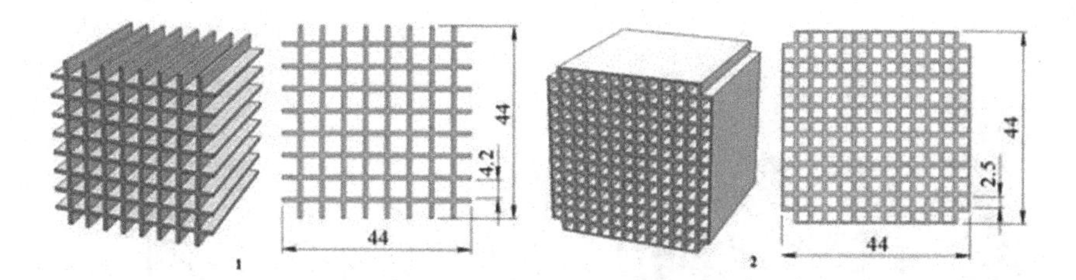


**FIGURE 4.11** External view of the monolith catalyst blocks with square channels, where the channel sizes are 4.2 mm (1) and 2.5 mm (2).

**FIGURE 4.12** External view of AP-64 catalyst granules and experimental cartridge.

The external view of the produced wire mesh cartridge is shown in Figure 4.13. The stainless-steel mesh with the external layer of the secondary support (alumina) was used for the cartridge production. Again, the Pt was supported in the same way as in GFC to minimize the difference of Pt state and its influence on the specific activity. Both flat and corrugated meshes in this monolith were catalytically active.

Properties of all comparative reference catalysts are summarized in Table 4.5.

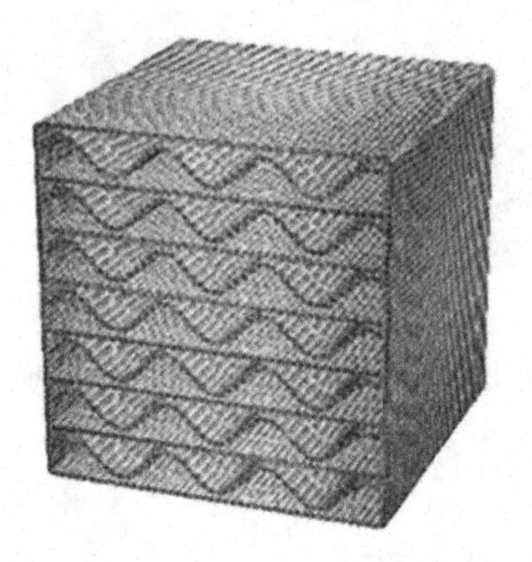

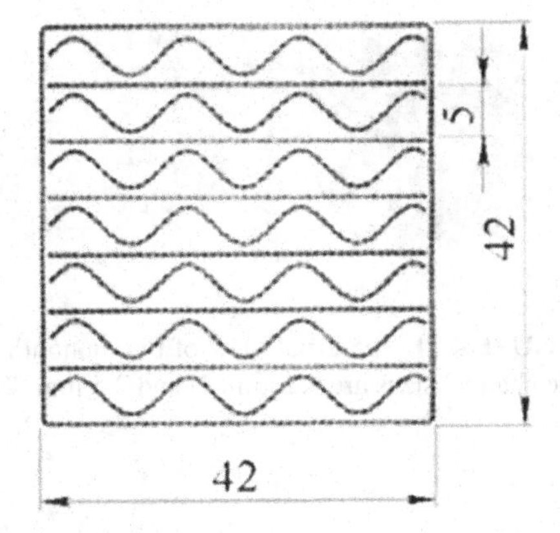


**FIGURE 4.13** External view of the wire-mesh block.

**TABLE 4.5**
**Properties of the Comparative Conventional Catalyst Samples**

| Parameter | Catalyst type | | | |
|---|---|---|---|---|
| | Ceramic monolith 4.2 mm | Ceramic monolith 2.5 mm | Granular packed bed | Wire mesh monolith |
| Catalyst mass in the cartridge, $m_{cat}$, g | 37.74 | 53.39 | 46.88 | 42.5 |
| Active component mass in the cartridge, $m_{ac}$, g | 0.03019 | 0.04271 | 0.03750 | 0.00798 |
| Porosity, $\varepsilon$ | 0.678 | 0.541 | 0.566 | 0.931 |
| Specific external surface area, $S_{sp}$, m$^{-1}$ | 646 | 865 | 756 | 554 |
| Equivalent pass diameter, $d_{eq}$, m | 0.0042 | 0.0025 | 0.030 | 0.0067 |

*Source*: Reprinted with permission from Zagoruiko et al. 2017.[4]

## 4.2. PRESSURE DROP IN GFC CARTRIDGES

### 4.2.1. Experimental Technique

Investigation of the hydraulic resistance of GFC cartridges and reference catalyst was performed at the experimental setup, which was schematically presented in Figure 4.14.[4,5]

Air was supplied into the main pipe of the setup by the means of ventilator. Air flow rate was controlled by the Mitsubishi frequency controller, allowing change in the electricity frequency in the ventilator engine with the step of 5 Hz. It provided accurate control of air velocity inside the pipe in the range from 0 to 30 m/sec. Pressure (depression) inside the pipe was measured by DMP-31 sensor with the basic measurement error of 0.2%, the results were transferred in digital form to the processing device Thermodat 17E3 for the indication and storage.

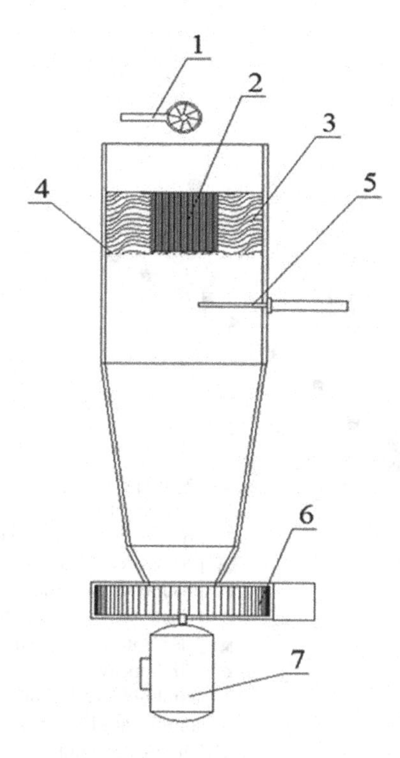


**FIGURE 4.14** Scheme of the experimental setup for the pressure drop studies: (1) handheld thermoanemometer, (2) catalytic cartridge, (3) sealing, (4), supporting grid, (5) pressure and temperature sensor, (6) ventilator, (7) electric engine.

The air velocity in the pipe was measured by the digital thermoanemometer-pyrometer DT-620; the relative measurement error in the temperature range from 0 to 60°C was not exceeding ±3%.

In the experiments, the cartridge was placed on the supporting grid above the pressure sensor. The clearances between the block and pipe walls were sealed to avoid air bypassing.

### 4.2.2. Experimental Results

The summary of the pressure drop tests is presented in Figure 4.15.

The following conclusions can be made on the base of the obtained results:

- The specific pressure drop of cartridges significantly depends on the channel size—pressure drop decreases with increasing distance between GFC layers.
- The hydraulic resistance of the cartridges does not depend on the structural type of GFC fabric (dense sateen or light openwork) or on the external shape of the cartridge.

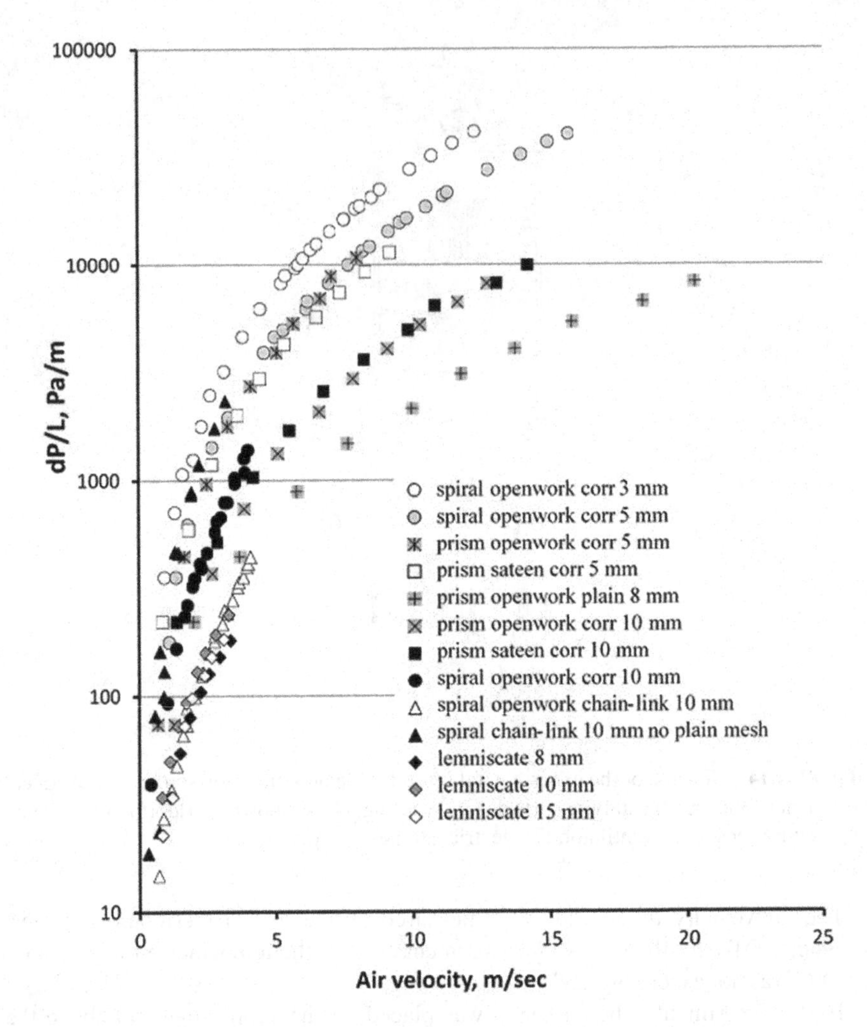


**FIGURE 4.15** Experimental dependence of the specific hydraulic resistance of GFC cartridges (per unit cartridge length) on air velocity at 20°C. Spiral/prism refers to the external cartridge shape; openwork/sateen to the GFC fabric type; corr/plain/chain-link relates to the type of structuring elements.

- Pressure drop decreases significantly if the volumetric chain-link net is applied instead of the corrugated mesh, and further decrease of the hydraulic resistance can be achieved for cartridges with plain structuring meshes, with the absence of the volumetric structuring elements inside the channels.
- Pressure drop significantly increases when the plain structuring meshes between GFC fabrics are not applied due to the destruction of the cartridge's regular geometry.

The dependence of the pressure drop in cartridges on the fluid velocity turned out to be close to quadratic. It is interesting that in the fixed beds of granular catalysts the order dependence of $\Delta P$ on the flow speed has a power order varying from ~1 (for laminar fluid) up to ~2 (for well-developed turbulent flow).[6] For the structured monolith catalysts, where the laminar regime is realized even at high flow velocity, this power order is also close to the unit.[7] Therefore, the structured GFC-based cartridges with a value closer to 2 provide the efficient turbulization of the reaction flow even at low and average velocities.

According to the known data,[8] the following equation is the most suitable for the calculation of pressure drop in various catalytic beds:

$$\Delta P = \xi \frac{\rho v^2}{2} \frac{S_{SP}}{\varepsilon^3} L, (Pa) \tag{4.18}$$

where $\Delta P$ is the pressure drop (Pa); $\xi$ is the hydraulic resistance coefficient (dimensionless); $\rho$ is the flow density (kg/m$^3$); $v$ is the fluid velocity (m/sec); $S_{SP}$ is the specific overall surface area of the cartridge (m$^{-1}$); $\varepsilon$ is the cartridge void fraction (dimensionless); and $L$ is the cartridge length in the direction of the flow (m).

Figure 4.16 shows the dependence of the hydraulic resistance coefficient $\xi$ on Reynolds number, which was calculated in this case as follows:

$$Re = \frac{4v\rho}{S_{SP}\mu} \tag{4.19}$$

where $\mu$ is the flow viscosity (Pa*sec).

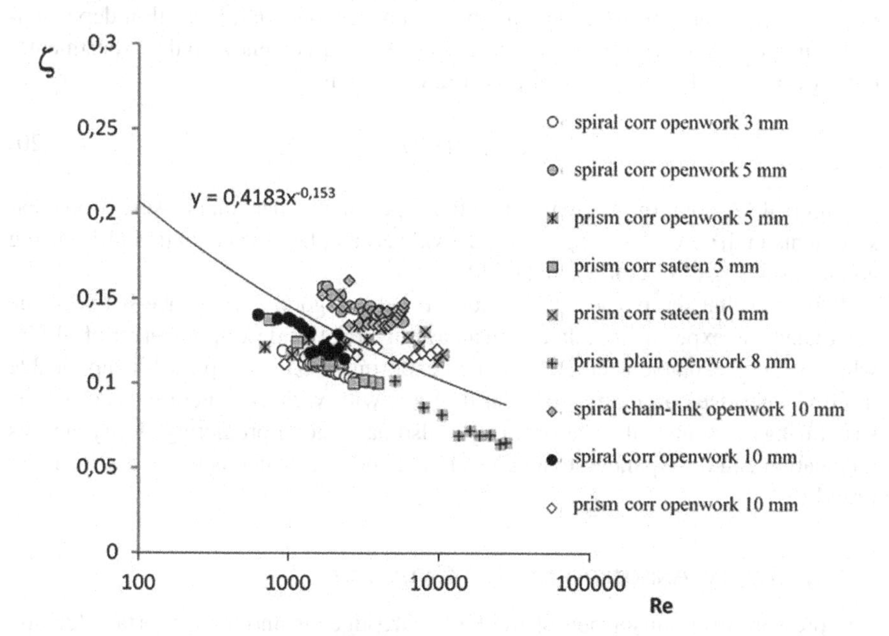


**FIGURE 4.16** Dependence of the hydraulic resistance coefficient $\xi$ on Reynolds number (points are experimental data, line is approximation by the power function).

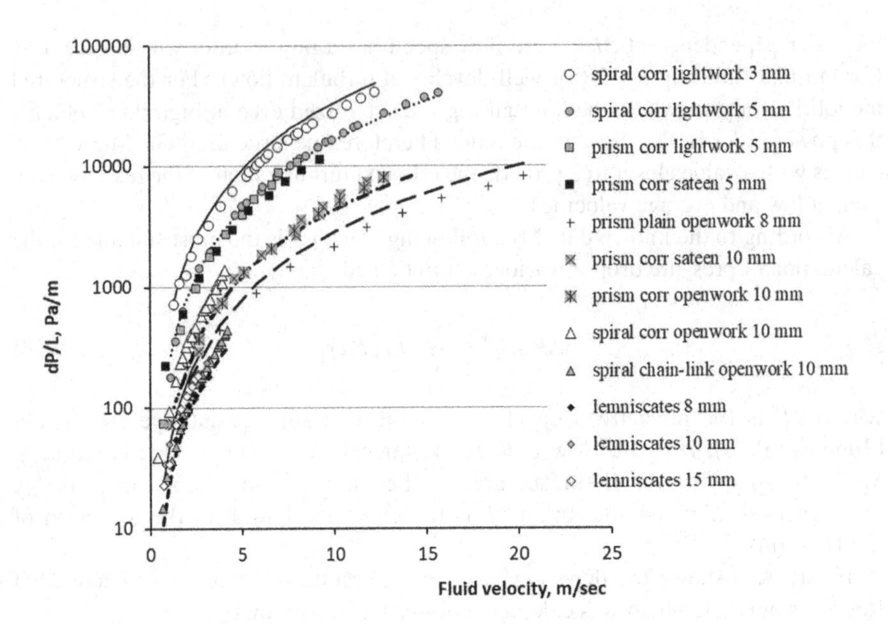


**FIGURE 4.17** Comparison of the experimental and calculated specific pressure drop in GFC cartridges (points are experimental data, line is calculations).

Value of the friction factor $\zeta$ was formulated as an empirical function of Reynolds number. Different types of functions (power law, exponential, logarithm dependencies, function $\zeta = A/Re + B$) were studied. The best agreement with the experimental data was achieved with the use of power law function:[9,10]

$$\xi = 0.418 Re^{-0.153} \tag{4.20}$$

Figure 4.17 shows the comparison of the experimental data on pressure drop measurements in different cartridges with the values calculated via equation (4.18) using equations (4.13), (4.14), (4.19), and (4.20).

It is seen that the proposed equation provides a good congruence between the calculated and experimental data with an average relative description error of ~16 %, which is an acceptable accuracy for such experiments. This approach is applicable to GFC cartridges of a very different design—with various structure, types of the structuring elements, GFC fabrics, etc. It also has a good predicting ability, and its application can be expanded to a new cartridge design, which is not experimentally tested yet.

### 4.2.3. Partial Anisotropy of GFC Cartridges

The pressure drop anisotropy of the GFC cartridges is another important feature, which can significantly influence the operation of the GFC-based reactors.[11] Here anisotropy is understood as a relative difference in pressure drop values, which is

**FIGURE 4.18** Basic directions of flow movement during the flow anisotropy measurements.

*Source*: Reprinted with permission from Zagoruiko et al. 2014.[11]

observed when the flow is moving through the cartridges in different directions. While the fixed bed of the conventional catalyst pellets is usually completely isotropic, and the conventional monolith catalysts are completely anisotropic, the fiberglass cartridges with the structuring gauzes are intuitively considered as a partially isotropic, as long as both the catalyst fabric and structuring metal gauze are permeable for the flow. The aim of this study was to quantify the level of isotropy and estimate whether it should be taken into account during the development and optimization of the reactor design.

In addition to the main gas movement direction (direction (a) in Figure 4.18) studied above (when gas moves along the gas passages formed by GFC planes and corrugation channels), two other basic directions were taken into consideration: (b) with gas moving across the gas passages, but along the GFC planes, and (c) with gas moving across both the gas passages and catalyst planes.

The experiments were performed with a prismatic cartridge of 5 mm step between GFC layers. The corrugated and flat stainless-steel gauzes with a cell size of 1 × 1 mm and wire thickness of 0.2 mm were used as the structuring elements. The experimental results are demonstrated in Figure 4.19.

While the pressure drop in case of flow movement across the passages and layers is much higher ($\xi_c/\xi_a \approx 80$) than that for the flow movement along the passages, the pressure drop relation between flow movement along the passages and across passages though simultaneously along the fabric layers is much more moderate ($\xi_b/\xi_a \approx 4.5$).

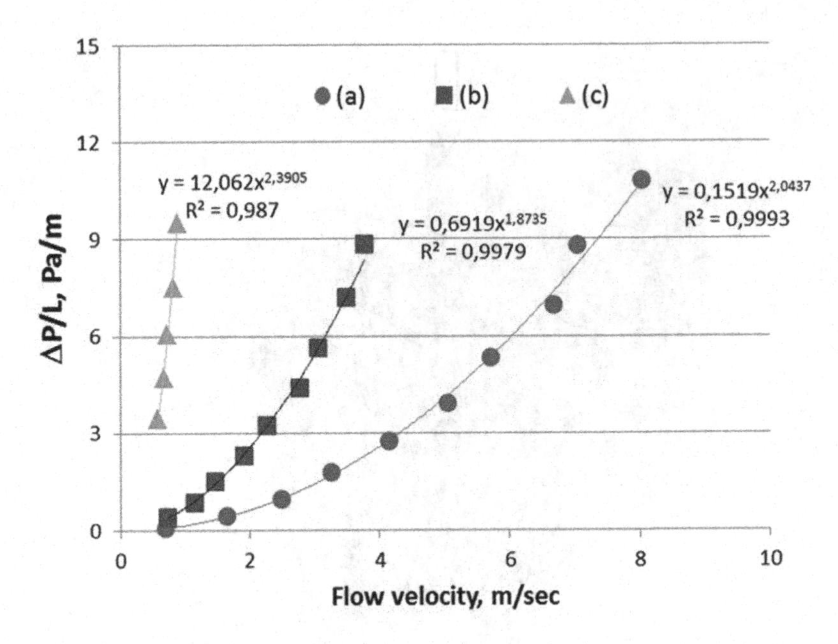


**FIGURE 4.19** The comparative pressure drop of the GFC cartridge for the different directions (a, b and c, as shown in Figure 4.18) of the gas flow.

*Source*: Reprinted with permission from Zagoruiko et al. 2014.[11]

Therefore, we may conclude that the cartridge is partially isotropic in direction (b) and completely anisotropic in direction (c).

This result means that the GFC cartridge may provide some internal flow redistribution in the direction (b) in case of non-uniform flow distribution at the cartridge inlet, unlike the conventional structured catalyst monoliths, where gas passages have no internal connections, thus such redistribution cannot be achieved. It is possible to state that GFC cartridges may combine the advantages of the conventional catalyst types in case of their proper positioning in the reactor: the low pressure drop, which is typical for the monolith catalysts, with a flow self-redistribution, improving the uniformity of the reaction flow in the catalyst bed, common for the fixed beds of granular catalysts.

The value of the pressure drops for the mentioned direction in respect to the main fluid direction (~4.5) is less than magnitude order; therefore, it should be taken into account at the reactor design, in particular, in CFD modeling of the reactors with the structured cartridges using fiber-glass catalysts.

## 4.3. INVESTIGATION OF MASS TRANSFER IN GFC-BASED CARTRIDGES

### 4.3.1. Experimental Technique

The test reaction of toluene deep oxidation in the air was selected for the mass transfer studies. Experiments were performed in the flow reactor of square-shaped cross-section made of stainless steel. The reactor was equipped with the two-zone heating system and the inert bed material for the inlet mixture preheating and uniform distribution across the catalyst bed. The temperature in the reactor was controlled by three thermocouples located inside the catalyst bed and its inlet and outlet.

Reaction mixture was prepared by the mixing of the air flow with the toluene-helium mixture flow from the saturator. The flow rates of the both streams were controlled by flow-mass controllers.

The composition of the inlet and outlet mixtures was analyzed by TSVET-500 GC with the flame-ionization detector. The measurements at each experimental point were repeated a few times until the establishment of the steady-state catalyst performance. The measurement of conversions was accompanied by the measurements of the pressure drop in the catalyst bed.

All catalysts were charged into the reactor in the form of cubic cartridge with the size of 44 × 44 × 44 mm (see Chapter 4.1). Such form was appropriate to provide the exact reproduction of the internal geometry of commercial-size packing for all tested catalysts.

The inlet reactor section included the fixed bed of steel 3 mm balls with the height of 180 mm to ensure the isothermal reaction volume and uniform distribution of the reaction flow across reactor sequence. Temperature distribution across the GFC bed was measured by moving the thermocouple in the bed channels. The measurement showed that maximum difference in the cartridge temperature did not exceed 1°C; this accuracy is quite sufficient for the correct attribution of temperature to the experimental results. The quality of flow distribution was initially rather high; the pressure difference was insufficient for the quantitative measurements. Therefore, this test was performed using visual monitoring of smoky stream supply applying the simultaneous tracing of the flow distribution with laser. The obtained results proved the uniform distribution of the flow across the whole square reactor sequence; no deadlock zones or jet streams were observed (see Figure 4.20).

Toluene concentration in the produced mixture was kept at the level of ~100 ppm. On the one hand, such a concentration is low enough to minimize the heat emission from the oxidation reaction, and thus provides the minimum temperature gradient in the catalyst bed. On the other hand, it is high enough to ensure accurate measurement of the toluene conversion in the reaction.

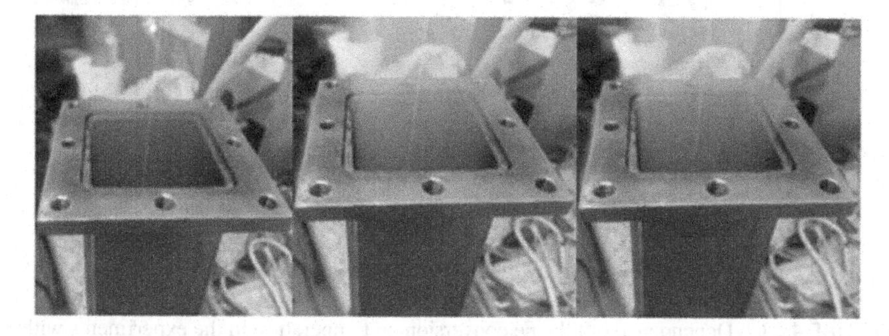

**FIGURE 4.20** Visual monitoring of the flow distribution in the experimental reactor sequence.

The first experimental series included measurement of the toluene conversion at various temperatures in the range of 100–500°C. To provide a more or less equal conditions for the comparison and to minimize the influence of substantially different mass of GFCs in different cartridges, these experiments were performed at a constant specific flow rate of the reaction mixture per unit mass of catalyst (2.5 L/min per 1 g of catalyst) with all types of catalytic structures, except for high-density catalysts (ceramic and wire-mesh monoliths, granular catalyst), tested under a maximum possible flow rate of 70 L/min.

The second experimental series included a variation of the flow rate at constant temperature. The basic temperature was set at 350°C, as this value was relevant to the realization of the deep external diffusion limitations regime for all tested catalysts, according to the analysis of the apparent reaction kinetics described below.

### 4.3.2. Experimental Results

The results of the experiments at constant flow rate and variable temperature are presented in Figure 4.21.

It is seen that the apparent activity per unit mass of the active component is practically equal in all cartridges in the temperature range up to ~180°C. It means that at such temperature the reaction rate is kinetically controlled. At higher temperatures the difference between the performance of different cartridges becomes substantial due to the different mass transfer limitations.

Figure 4.22 shows the dependence of toluene conversion on the reaction mixture flow rate at constant temperature of 350°C. In this case the reaction occurs in the

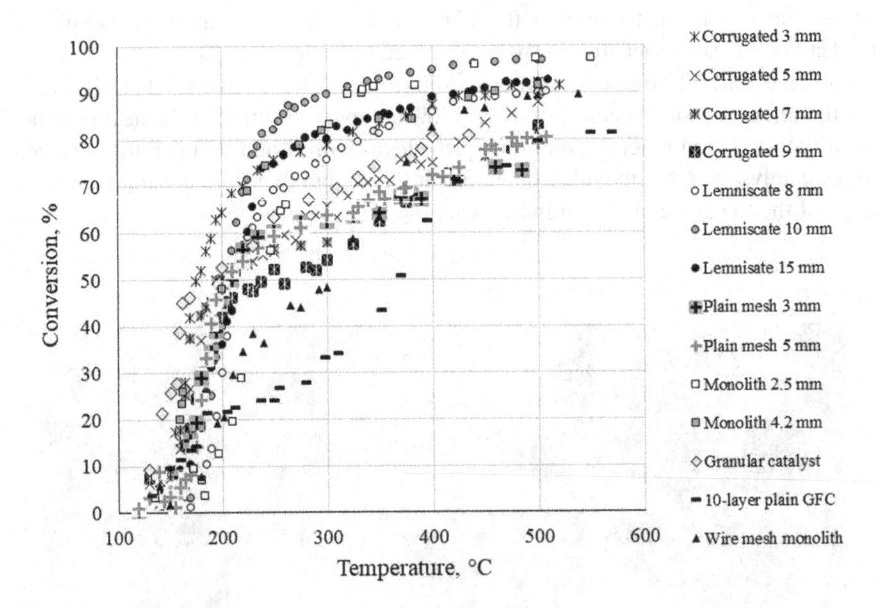


**FIGURE 4.21** Dependence of toluene conversion on temperature in the experiments with a constant flow rate (2.5 L/min per 1 g of catalyst).

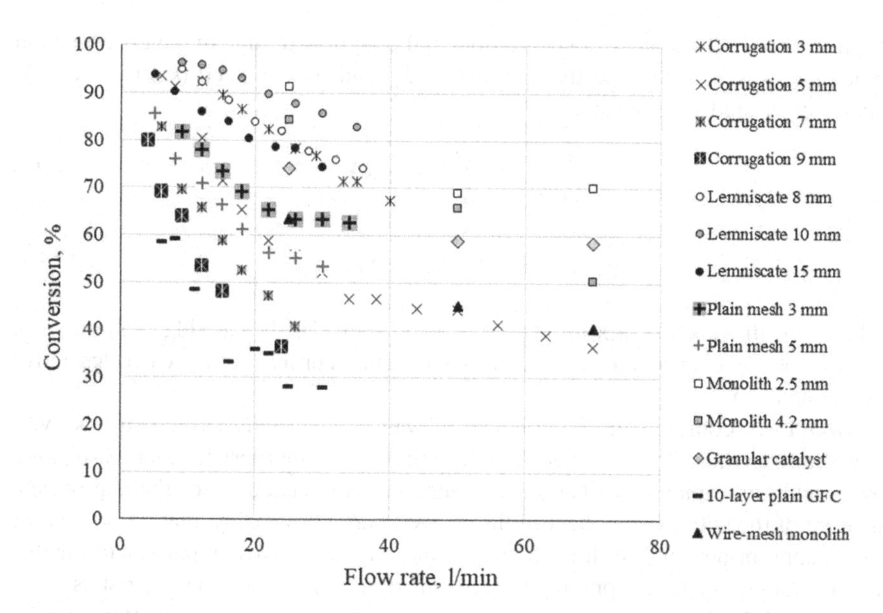


**FIGURE 4.22** Comparative dependence of toluene conversion on reaction mixture flow rate at 350°C.

regime, which is completely controlled by the external diffusion limitations, and changes in the apparent activity are caused by the change of mass transfer efficiency for various catalyst shapes and fluid velocities.

As expected, the observed conversion decreases with the increase of the flow rate (due to corresponding decrease of the residence time) for all samples, though the slope of this decrease becomes less with the increase of gas velocity due to the improvement of mass transfer in the catalyst beds. The apparent volumetric activity quite logically decreases for each type of the catalyst packing with the increase of the gas passage height due to a lower specific area of the catalyst surface. In case of GFC structures, such a decrease is additionally caused by a lower amount of GFC in cartridges with larger passages. The only exception here is the lemniscate structures—actually, the different lemniscate GFCs samples have different weaving structure and the connection between lemniscate layer height and its specific surface area is indirect (see Table 4.2).

As seen from Figure 4.22, the highest volumetric-apparent activity among all tested types of the catalyst structures is demonstrated by the lemniscate GFCs and ceramic monoliths, a slightly lower activity is observed for GFC cartridges of both flat and corrugated structure, as well as in the packed bed of granular catalysts. The lowest activity is observed for the multilayered GFC stack.

As Tables 4.1–4.5 show, the amount of catalyst (by mass) in all GFC cartridges is substantially smaller (by few times) than that for the conventional catalysts, despite the equal volume. We applied the individual apparent reaction rates to provide the objective comparison between the samples, both related to the catalyst bed volume and to the active component mass. Proceeding from the first-order kinetic equation (in reference to the toluene concentration) and plug flow regime in the experimental reactor, it was better to use the volumetric ($k_v$) and mass-related ($k_m$) kinetic constants, computed as follows:

$$k_V = -\frac{Q}{V}\ln(1-x) \tag{4.21}$$

$$k_m = -\frac{Q}{m}\ln(1-x) \tag{4.22}$$

where $Q$ is the reaction mixture flow rate (st.m$^3$/sec), $V$ is the cartridge volume (m$^3$), $x$ is the toluene conversion, and $m$ is the catalyst mass or mass of the cartridge active component (g).

Both these constants have a practical relevance. The catalyst volumetric activity characterizes the appropriate bed volume, and thus the appropriate reactor size, and, as a result, its capital cost. The mass-related activity characterizes the appropriate amount of the active component in the catalyst bed, allowing estimation of the cost of the appropriate catalyst. It is possible to make an objective comparison of the different catalytic systems, applying the combination of these two characteristics.

Figure 4.23 demonstrates the temperature dependence of the apparent reaction constant related to the mass of the active component.

In particular, it is seen that the different GFC structures have a different activity depending on the cartridge geometry. As long as the GFC itself is present in all these beds (except lemniscates) in the same form—GFC fabric of the same thickness—the

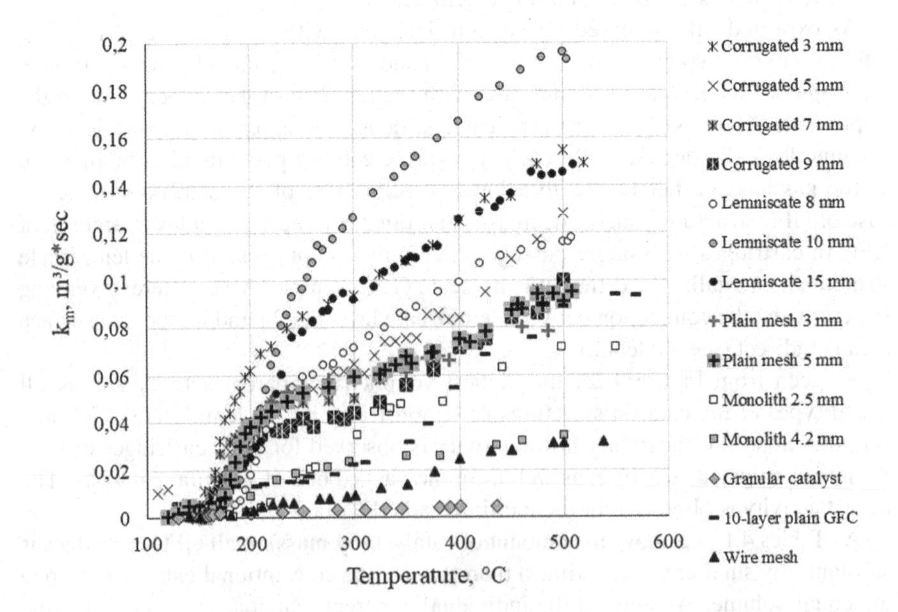


**FIGURE 4.23** Comparative dependence of the specific apparent reaction rate constant related to the unit mass of the active component on temperature.

internal diffusion limitations are the same; hence, all the difference at temperatures exceeding 200°C is caused only by the external diffusion limitations.

The highest apparent activity is demonstrated by the lemniscate GFCs. It is caused by the minimum overlapping of the threads; they are also not exposed to the shielding effect of the metal structuring elements, making the threads much more accessible for the reactants. Structure of the lemniscate elements provides the efficient flow turbulization, thus favorably affecting the external mass transfer. Moreover, under the influence of the moving fluid the intra-thread convective flows may appear, increasing both the internal and external mass transfer. All of this ensures higher overall mass transfer intensity in the lemniscate structures.

The next group of GFCs in the apparent activity are the cartridges with the corrugated and plain metal meshes. Their activity evidently and expectedly decreases with the increase of height between GFC layers. This is caused by the extension of the equivalent passage diameter and corresponding worsening of the mass transfer conditions.

GFC cartridges with the plain structuring meshes have lower activity than the similar cartridges with the corrugated structuring meshes. It can be possibly caused by the turbulizing effect of the corrugation, resulting in the mass transfer intensification. As expected, the most active cartridge for this type is the one with a lower passage height.

The worst performance of all GFC structures is demonstrated by the multilayered stack with the propagative flow. Theoretically this packing seemed good enough due to the possible presence of forced convective flows inside the threads, which may significantly increase the actual area of the catalyst washable surface, and thus lead to a very high efficiency of the internal and external mass transfer. But in fact, the observed activity was much lower than expected.

In our opinion, it is explained by the specific relative allocation of GFC threads in this packing.[12] The possible variants of such pattern are shown in Figure 4.24, they can be optionally divided into the checkerboard, asymmetrical, and corridor types.

The checkerboard pattern, characterized with a maximum contact of the fluid with the thread external surface and stimulation of the convective flow inside the threads, is obviously the most efficient type of structure in terms of overall mass transfer intensity. The corridor pattern is less efficient, as a large amount of the internal and external surfaces of the GFC threads become unavailable for the convective flow. Asymmetric structures are intermediate between them, with some intermediate activity.

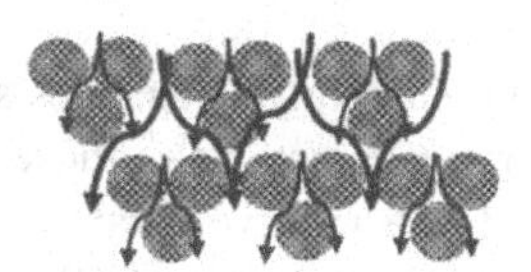
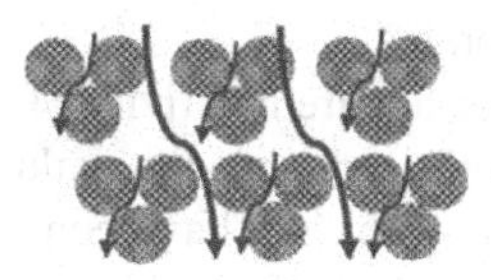
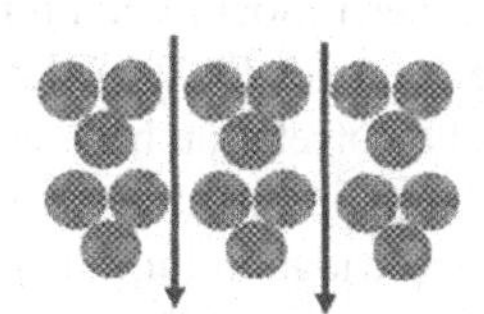

**FIGURE 4.24** Checkerboard (on the left), asymmetrical (in the center), and corridor (on the right) packing of GFC threads in the adjacent fabric layers. Arrows show the possible directions of the reaction flow movement.

Unfortunately, at present no methods are available for the purposeful creation of multilayered GFC packing with a purely checkerboard-type structure. Moreover, experimental data[12] have shown that GFC layer superposition can be self-transformed under the influence of the reaction fluid into patterns with a minimum possible pressure drop (i.e., even if not purely corridor packing, but at least asymmetric type). In other words, the self-structuration of the multilayered packing should be directed toward the decrease of the mass transfer efficiency.

In fact, the presented image of the possible pattern is too simplified except for the presented plain structures; it is possible to expect the formation of more complicated three-dimensional non-uniformities of various shapes and scales with the increased permeability for the reaction fluid. This opinion is based on our experience with multilayered GFC packing; they typically have a quite low level of results reproducibility in very different experiments.

Summarizing, the structured GFC cartridges with the axial flow are more efficient in terms of mass transfer and provide much better reproducibility (and, therefore, the operation reliability) than multilayered GFC packing with the propagative fluid.

All the considered catalyst types showed lower specific activity in comparison with GFCs. In the monoliths with the smooth channels it could be attributed to the laminar regime of flow movement leading to a low external mass transfer efficiency. However, to some extent it can be also connected with the internal diffusion limitations inside the pores of the monolith alumina support.

The visible negative influence of the internal mass transfer limitations is observed in case of the fixed bed of granular catalyst, characterized with the maximum thickness of catalytic material and, simultaneously, with the minimum specific activity among all tested catalysts.

The wire-mesh block demonstrates the highest apparent activity among the conventional catalysts. It can be connected both with the systematic asperities of wire meshes, turbulizing the flow and improving the external mass transfer, and with the minimum internal mass transfer limitations due to the negligible thickness of the active layer on the wire surface.

The described regularities are reproduced in general in case of the specific rate, calculated per unit mass of the catalyst (Figure 4.25).

The only visible difference is the relative improvement of the specific apparent rate constant for the granular catalyst, which becomes competitive with other types of conventional catalysts. This is owing to a higher Pt loading (~0.6 % mass) than in all other studied systems (0.04–0.07 %). To the contrary, the performance of the wire-mesh block, containing the minimum platinum amount compared to all the other samples (0.02%), deteriorates.

The correlation between the apparent activity of different catalysts changes significantly in the transfer from the mass specific values of rate constants to those for unit cartridge volume (Figure 4.26). As has been noted, the relative catalyst bed density (mass of the catalyst in the unit cartridge volume) in the case of GFCs is significantly lower than for other catalysts. Respectively, in the comparison on the base of volume-related rate constants the GFCs seem less advantageous, while the apparent performance of monoliths and granular beds looks better.

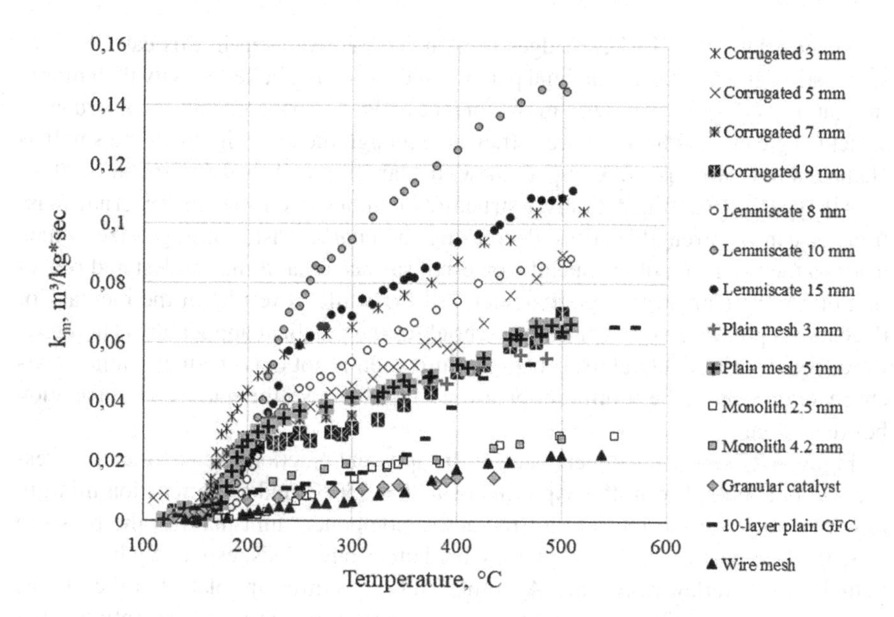


**FIGURE 4.25** Comparative dependence of the specific apparent reaction rate constant related to the unit mass of the catalyst on temperature.

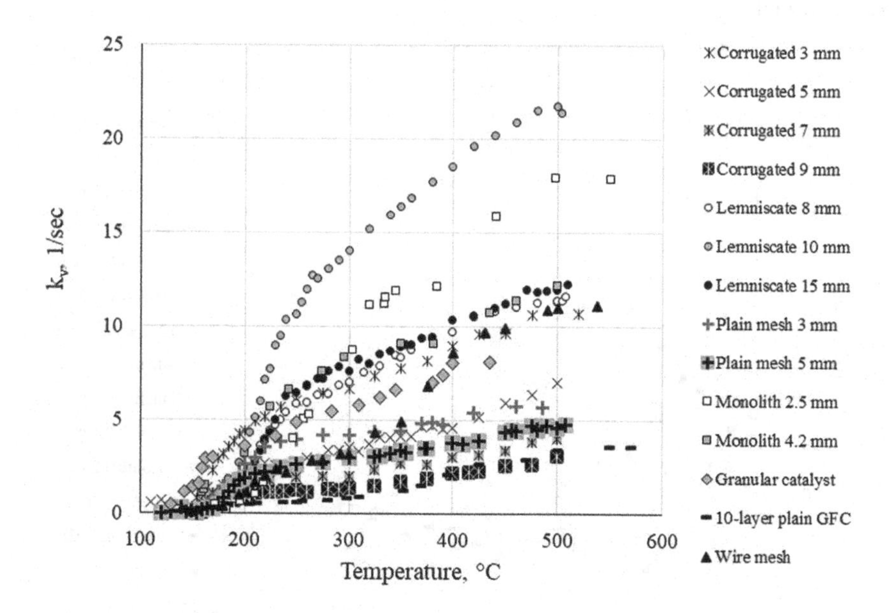


**FIGURE 4.26** Comparative dependence of the specific apparent reaction rate constant related to the unit cartridge volume on temperature.

Nevertheless, the GFC cartridges remain competitive even in this case. Lemniscates still demonstrate a beneficial performance, showing better activity than monolith catalysts despite the lower mass content of Pt. Cartridges with the corrugated structuring elements have a lower efficiency, though the cartridge with the smallest channel size (3 mm) is still competitive with a larger monolith and granular catalyst.

All types of examined catalyst structures demonstrated that the external mass transfer can be strengthened by decreasing the characteristic passage size, which leads to the extension of the specific external surface area of the catalyst and reduction of the equivalent passage diameter. All the while, it results in the increase of the catalyst pressure drop, which is essential for the practical application. Therefore, the adequate criteria of catalyst shape optimization is not only the maximum apparent reaction rate or the minimum pressure drop, but also the quality of correlation between them.

Figure 4.27 shows the dependence of the apparent reaction rates on the unit pressure drop, obtained from the experiments at T = 350°C under the reaction mixture flow rate variation. In the ratio of the active component unit mass to the pressure drop the best performance is shown by the lemniscate GFCs, especially by the one with 15 mm interlayer distance. All other GFC structures are placed at the plot as a dense group, all with good activity/pressure drop ratio. The only exception is the multilayer GFC stack, which has rather high pressure drop with the propagative type of flow through dense microfibrous medium. Monolith, wire-mesh, and granular catalysts have a lower efficiency as a result of lower values of the apparent mass–related rate constants.

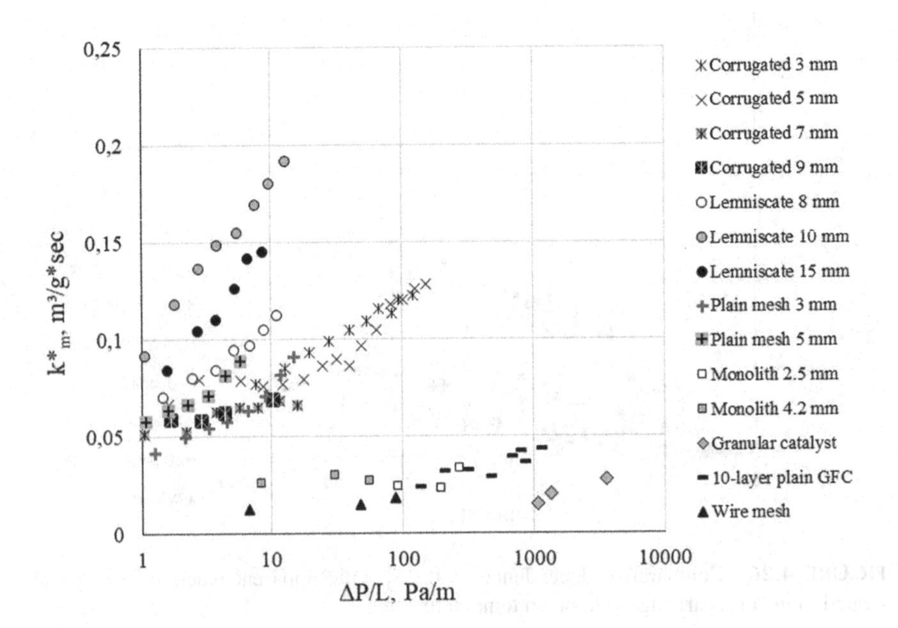


**FIGURE 4.27** Comparative dependence of the apparent reaction rate constant per unit mass of Pt on the specific pressure drop at 350°C.

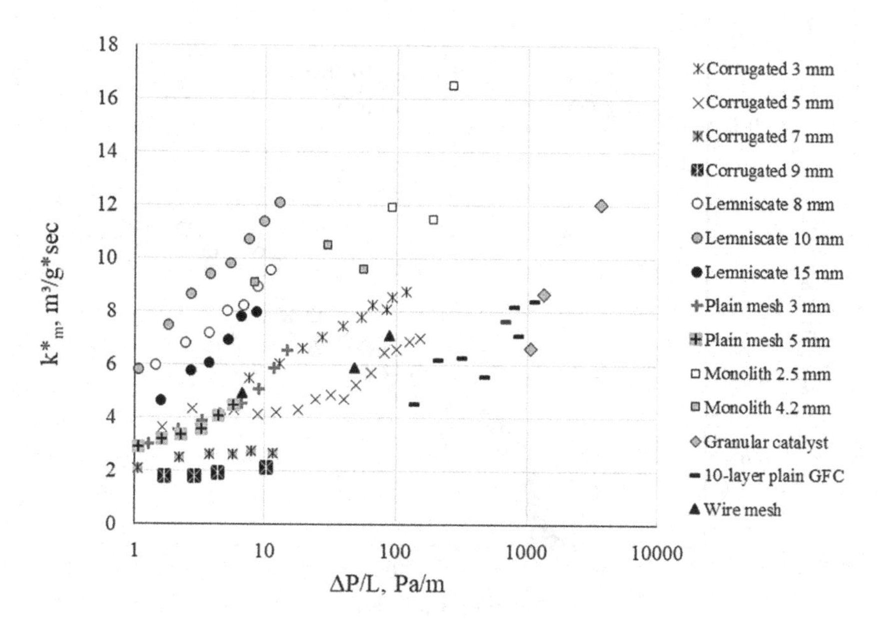


**FIGURE 4.28** Comparative dependence of the apparent reaction rate constant per unit mass of Pt on the specific pressure drop at 350°C.

According to the ratio between the volume-related rate constant and a specific pressure drop (Figure 4.28), the picture looks different. Due to a higher relative density, the monolith and wire-mesh catalysts move closer to the dense GFC group. Though the fixed bed of the granular catalyst has a volumetric activity of the same order, it remains in the worst position due to its high pressure drop.

Summarizing, the proposed GFC-based structures show excellent performance, being competitive with the best-known conventional catalyst shapes in relation to the apparent activity per unit volume, at the same time significantly exceling them in the activity-related unit mass of the catalyst or unit mass of the active component. In addition, they demonstrate the best ratio between the apparent activity and specific pressure drop. The best performance is demonstrated by the lemniscate GFCs among all the considered GFC structures.

### 4.3.3. Intrinsic Kinetics of the Toluene Oxidation Pt/GFCs

GFC cartridges with the corrugated structuring meshes were used to analyze the intrinsic kinetics of the toluene oxidation. The results of the kinetic analysis based on the experiments with a temperature variation are given in Figure 4.29.

It appears that the apparent activity per unit mass of the active component is almost the same for all cartridges in the temperature range up to ~180°C (Figure 4.29a). It shows that the reaction rate is kinetically regulated at these temperatures; the discrepancy between the performance of different cartridges becomes meaningful at a higher temperature due to various mass transfer limitations.

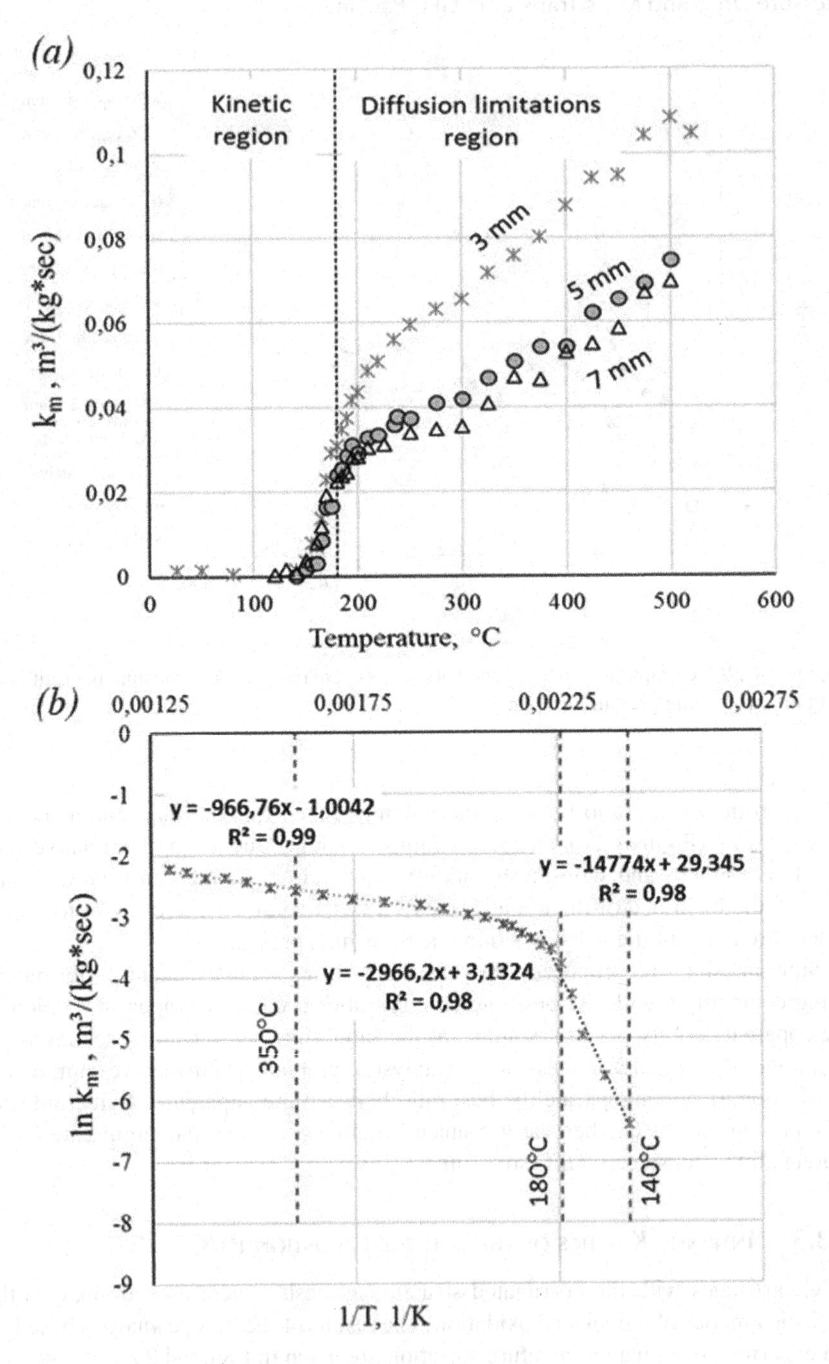


**FIGURE 4.29** Comparative dependence of the apparent rate constant of the toluene oxidation per unit catalyst mass on temperature (a) in the GFC cartridges with the corrugated structured mesh elements (corrugation height of 3, 5, 7 mm) and (b) dependence of the apparent reaction rate constant logarithm on the inverse temperature.

*Source*: Reprinted with permission from Zagoruiko et al. 2017.[4]

Therefore, the experimental data in the temperature range of 140–180°C were taken to analyze the intrinsic kinetics of the toluene oxidation. The temperature relation to the apparent mass-related kinetic constant is specified in Arrhenius coordinates in Figure 4.29b.

It is observed that the rate constant logarithm values in the temperature range of 140–180°C can be summarized with a high accuracy by the linear function, which approved the idea about the first-order dependence of the reaction rate on the toluene concentration, as well as the assumption that there is no diffusion limitation in this temperature region. The apparent activation energy is equal to ~120 kJ/mole, which is similar to the characteristic values for the reactions of aromatic hydrocarbons deep oxidation at Pt-containing catalysts.[13,14]

In compliance with the approximation data, the apparent first-order kinetic constant for the reaction of toluene deep oxidation in the GFC IC-12-S111 can be computed as follows:

$$k^* = 5.55 * 10^{12} \exp^{\frac{-14774}{T+273}} \rho_{GFC} \tag{4.23}$$

where $\rho_{GFC}$ is the catalyst mass in the GFC cartridge unit volume.

The apparent activation energy is lower at higher temperature, which is normal for the diffusion limitation of the reaction. The average activation energy in the temperature range of ~180–250°C is equal to ~25 kJ/mole, resulting from the possible influence either of internal and external diffusion limitations. In case the temperature is above 250°C, the apparent activation energy is almost constant and equal to ~8 kJ/mole, which is normal for the pure regime of the external mass transfer limitations. For this reason, the temperature of 350°C was adopted for the experimental studies of the external mass transfer.

### 4.3.4. Intra-Thread Diffusion Limitations

The GFC textile has relatively high internal porosity and a low typical thickness; hence, it is often concluded as being free of internal diffusion limitations. Our previous study[15] predicted the significant influence of intra-thread mass transfer resistance in the GFC of the thick (3 mm) lightweight structure. Concurrently, the possible influence of the internal mass transfer in the thinner textiles is not obvious and needed the additional consideration.

Thiele modulus values served as a basis to estimate the influence of the mass transfer limitations inside the GFC fabric. It can be calculated for the flat fabric and first-order reaction kinetics with the following equation:

$$\varphi = d\sqrt{\frac{k*}{D}} \tag{4.24}$$

As is known, the internal mass transfer limitations are to be considered, if $\varphi > 1$.[16] The dependence of Thiele modulus on temperature for GFC fabrics of a various thickness is shown in Figure 4.30.

The standard thickness of the glass-fiber fabric to produce the catalyst is 0.3–0.4 mm; thus, the mass transfer limitations for the described GFC are meaningful starting from

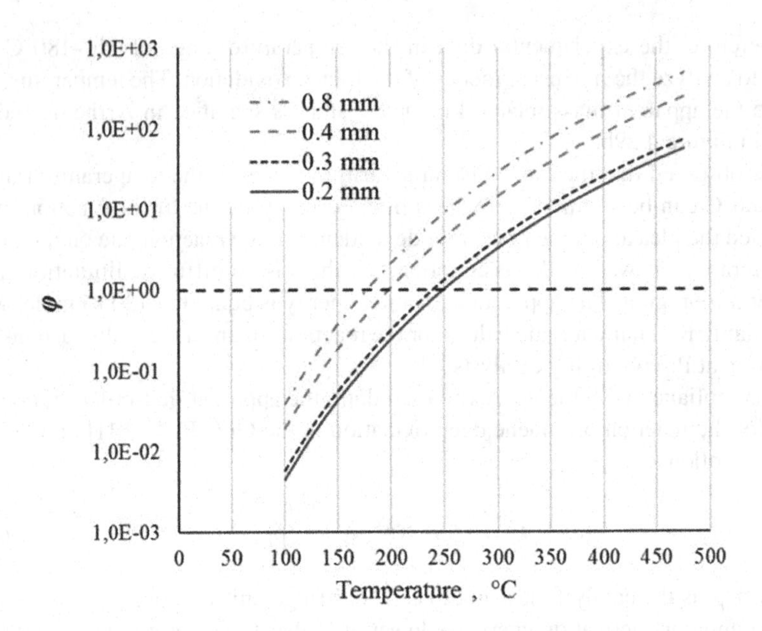


**FIGURE 4.30** Dependence of Thiele modulus on temperature for GFC fabrics of various thicknesses.

*Source*: Reprinted with permission from Zagoruiko et al. 2017.[4]

the temperature of ~200–240°C. Therefore, the intra-fabric diffusion limitations have to be considered in the mass transfer experiments performed at 350°C.

Account of such limitations was based on the efficiency factor calculation applying the equation for the flat (one-dimensional) catalytic textile and first-order reaction kinetics:[16]

$$\eta = \frac{\tanh \varphi}{\varphi} \tag{4.25}$$

## 4.3.5. On the Intra-Fiber Mass Transfer

The high activity and operation stability overseen in some previous works[17–19] for Pt-based GFC in the reaction of organic compound deep oxidation were related to the highly dispersed platinum particles present in the glass bulk under the glass-fiber surface, at the depth within 10–20 nm. The analogous subsurface palladium particles were offered as a highly selective catalyst for acetylene[17,20] hydrogenation. As long as the glass support is reckoned as non-porous matter, it is suggested that the reactant transportation from the fiber surface to subsurface metal particles happens through the diffusion in the glass bulk similar to liquid. This hypothesis is based on the idea that glass is a super-cooled liquid with a high viscosity.

This suggestion has been actively discussed for the last two decades. There is no direct evidence of the existence of such subsurface particles, so it is not fully convincing. Moreover, the existence of such particles is not apparent, and it is unclear, if they have a sufficient amount to provide the observed catalyst activity supposing they actually exist. Further, it is not possible to recognize the contribution of surface and subsurface particles into overall catalyst performance, as in all synthesized GFCs at least part of the catalytically active metal is present on the external fiber surface. Hence, any dispute on the catalytic properties of the subsurface species is not grounded on the direct experimental data.

Apart from the dispute on the existence of subsurface metal particles, it is worth paying special attention to the reactant transportation in the glass bulk. If the active subsurface species are available, and if their amount is big enough, the apparent reaction rates with a rather high value should be promoted by reactant transportation of the sufficient mass from fiber surface to these particles and backward diffusion of the reaction products. The potential existence of such flow can be assessed using the common approach on the base of Thiele modulus,[16] similar to that described in the previous chapter with the examination of a single GFC microfiber instead of GFC thread.

As this approach examines the species transportation inside the glass volume as diffusion in the liquid, it is possible to use the Stokes-Einstein equation for the diffusion in liquids to calculate the toluene diffusion coefficient:

$$D = \frac{k_B T}{6\pi\mu r} \tag{4.26}$$

The ambient and moderate temperatures (up to 300°C) are the most appropriate for GFC applications. Glass has a very high viscosity at such temperatures, especially the high-silica glasses used for GFC synthesis. It makes its correct measurement complicated; the data in the scientific literature are often specified for temperatures above 500°C and glasses with a decreased $SiO_2$ content and relatively low viscosity. The assessments[21,22] for lower temperatures are contradictory and vary from $10^{12}$ Pa*s up to $10^{21}$ to $10^{41}$ Pa*s; some discussions[23] are still ongoing even in respect to whether the glass at ambient temperature is a super-cooled liquid or an amorphous solid. In our opinion, the most reasonable and well considered estimation was given by Vannoni at al. 2011,[24] where the fused silica glass viscosity at ambient temperature was specified as $10^{17}$ to $10^{18}$ Pa*s. In the present study the more optimistic value $\mu = 10^{17}$ Pa*s was used.

Calculations were based on the model[25] considering the GFC fiber as a long straight cylinder with a length $L$ and radius $R$. Two geometrically different types of active component supports were assumed: the active component is supported on the external surface of the fiber or inside the fiber bulk layer with the thickness $h$.

The maximum thickness of the active layer for the second case was taken equal to 10 nm, the effective radius of the toluene molecule as 0.335 nm, and the average fiber radius as 2.5 microns. Kinetic data, described in Chapter 4.3.3, were used for the description of the specific reaction rate of toluene oxidation.

Calculations using the equation (4.26) show the very low value of toluene diffusion coefficient inside the glass bulk. In the temperature range from 0 to 300°C it is equal to $10^{29}$ to $10^{30}$ $m^2$/sec. This is by 23–24 orders of magnitude less than for toluene diffusion in the air at similar conditions and by 12–13 orders of magnitude less than evaluations made on the base of the experimental data, assuming the occurrence of oxidation reaction at subsurface active particles.[19] In fact, such low values of

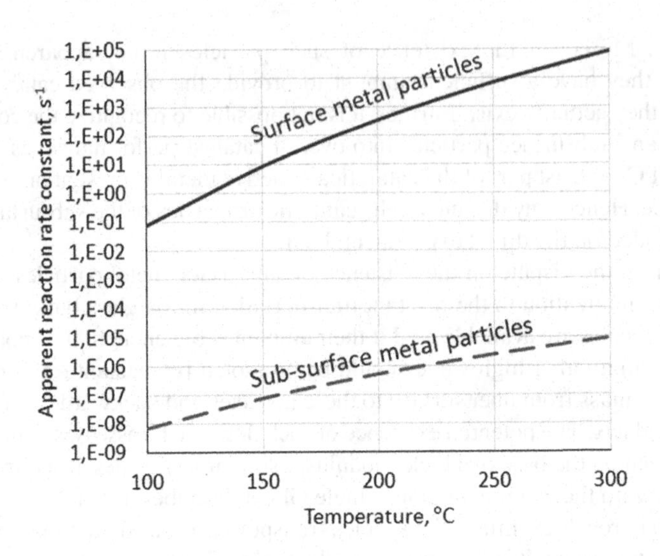


**FIGURE 4.31** Temperature dependence of the apparent rate constants.

the diffusion coefficient can be considered as practical impermeability of the glass for the diffusion of any molecules. The results of the comparative calculations of apparent kinetic constants for the two mentioned models are given in Figure 4.31.

It is obvious, that apparent activity of the catalyst with a subsurface location of active component is considerably lower than that for the catalyst with the surface active component: the difference is 7–10 orders of magnitude. In fact, the reaction occurrence at the subsurface particles seems to be impossible.

The additional calculations for the subsurface catalyst were made to exclude the influence of the possible false in model parameters, assuming that the glass viscosity of $10^{12}$ Pa*s is typical for low-silica glasses at higher temperature. Though it was observed the decrease of the difference with the surface catalyst, nevertheless, it is still very high (5–8 orders of magnitude). One more calculation was made for the catalyst with a thinner subsurface layer (1 nm instead of 10 nm); it caused an insignificant increase of the apparent activity. The calculation was also performed based on a hypothetic assumption that subsurface Pt clusters have a higher intrinsic activity than the surface ones. The multiplication of the initial $k_m^0$ value by 100 shows only one order improvement in the apparent reaction rate, having still very high (6–9 orders of magnitude) clearance in comparison with the surface Pt location. These results show that a big difference in the apparent reaction rates is in the wide range of model parameters and it is not the result of any inaccuracy or errors in the definition of parameter values.

Theoretically, you may assume that the liquid-phase diffusion approach is not proper for GFCs and that there may be some other way that ensures a higher actual internal diffusivity in the glass bulk. In fact, this assumption goes against the reality: glass vessels and pipelines are widely applied in chemistry and many other areas for the hermetic treatment and long-term storage of various liquids and gaseous substances. Obviously, it is actual fact that glass is impermeable even for a very light molecule like hydrogen or helium.

Therefore, the specific properties of GFCs, connected with their activity and selectivity, relate to the particular qualities of the active particles located on the surface of glass fibers and improved external mass transfer in GFC packing, rather than to any subsurface species with extraordinary properties.

### 4.3.6. External Diffusion Limitations in the GFC Cartridges

The studies of the external mass transfer in the GFC cartridges were based on the described experimental array, taken from the experiments with a constant temperature and variation of flow rate.

The apparent reaction rate constant in the external diffusion limitation regime for the first-order reaction kinetics is defined by the following equation:

$$\frac{1}{k^{**}} = \frac{1}{\beta S_{sp}} + \frac{1}{\eta k^{*}} \tag{4.27}$$

where $k^{**}$ is the apparent rate constant under diffusion limitations conditions (1/sec); $\beta$ is the mass transfer coefficient (m/sec); $S_{SP}$ is the specific external surface area of GFC (1/m); $\eta$ is the efficiency factor, accounting for intra-thread mass transfer (see equation 4.25); and $k^*$ is the intrinsic rate constant (1/sec). Therefore, the value of the mass transfer coefficient can be calculated based on the observed toluene conversion as follows:

$$\beta = \frac{1}{S_{sp}\left(\frac{1}{k^{**}} - \frac{1}{\eta k^{*}}\right)} \tag{4.28}$$

The dependence of mass transfer coefficient on the linear velocity of the reaction gas flow is plotted in Figure 4.32.

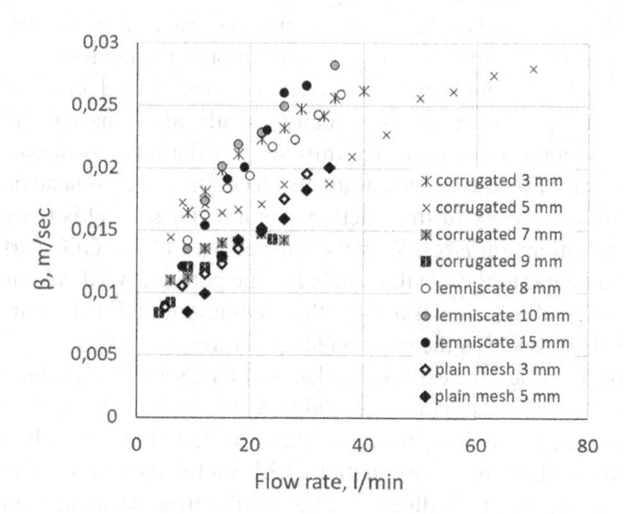


**FIGURE 4.32** The dependence of mass transfer coefficient on the linear velocity of the reaction gas flow.

It is seen that the mass transfer coefficient increases the flow speed for all types of cartridges. This is explained by flow turbulization under its increased velocity leading to corresponding intensification of the external mass transfer. You can also see the quite reasonable decrease of the mass transfer coefficient with the extension of the equivalent diameter of the passage in cartridges.

The obtained $\beta$ values were generalized in a form of the standard criterial equation:[9]

$$Sh = A\,\mathrm{Re}^n\, Sc^{1/3} \tag{4.29}$$

$$\mathrm{Re} = \frac{U d_{eq} \rho}{\mu} \tag{4.30}$$

$$Sh = \frac{\beta d_{eq}}{D} \tag{4.31}$$

$$Sc = \frac{\mu}{\rho D} \tag{4.32}$$

$$U = \frac{U_0 (T + 273)}{273 \varepsilon} \tag{4.33}$$

$$U_0 = \frac{Q}{S_R} \tag{4.34}$$

$$d_{eq} = \frac{4\varepsilon}{S_{SP}} \tag{4.35}$$

where $Sh$ is the Sherwood number (sometimes also referred to in the literature as the mass-related Nusselt number, $Nu_m$), a dimensionless criterion characterizing the efficiency of mass transfer; $A$ is the dimensionless empiric parameter; $Re$ and $Sc$ are the Reynolds and Schmidt numbers, respectively; $U$ is the gas velocity in the cartridge channel (m/sec); $U_0$ is the initial gas velocity, calculated for the standard conditions and full cross section of the cartridge (m/sec); $\rho$ is the real gas density in reaction conditions (kg/m$^3$); $D$ is the molecular diffusion coefficient of toluene in air (m$^2$/sec); $\mu$ is the dynamic viscosity of the reaction mixture (Pa*sec); $Q$ is the gas flow rate at standard conditions (m$^3$/sec); $S_R$ is the surface area of the GFC cartridge cross-section perpendicular to the gas flow (m$^2$); $\varepsilon$ is the porosity (void volume fraction) of the GFC cartridge (dimensionless); $S_{sp}$ is the specific external surface area of GFC in the cartridge (1/m); and $T$ is the reaction temperature (°C).

As mentioned earlier, the external surfaces of GFCs and corrugated wire meshes form the specific surface area in the cartridges with corrugated structuring elements. The equivalent hydraulic diameter of the pass for $Re$ number calculations in equation (4.30) was defined from the equation (4.35) on the basis of the overall specific surface area to calculate the influence of these structuring elements on flow turbulization. At the same time, the $Sh$ number values for this cartridge type were computed without considering the presence of the corrugated meshes, as long as in this case the

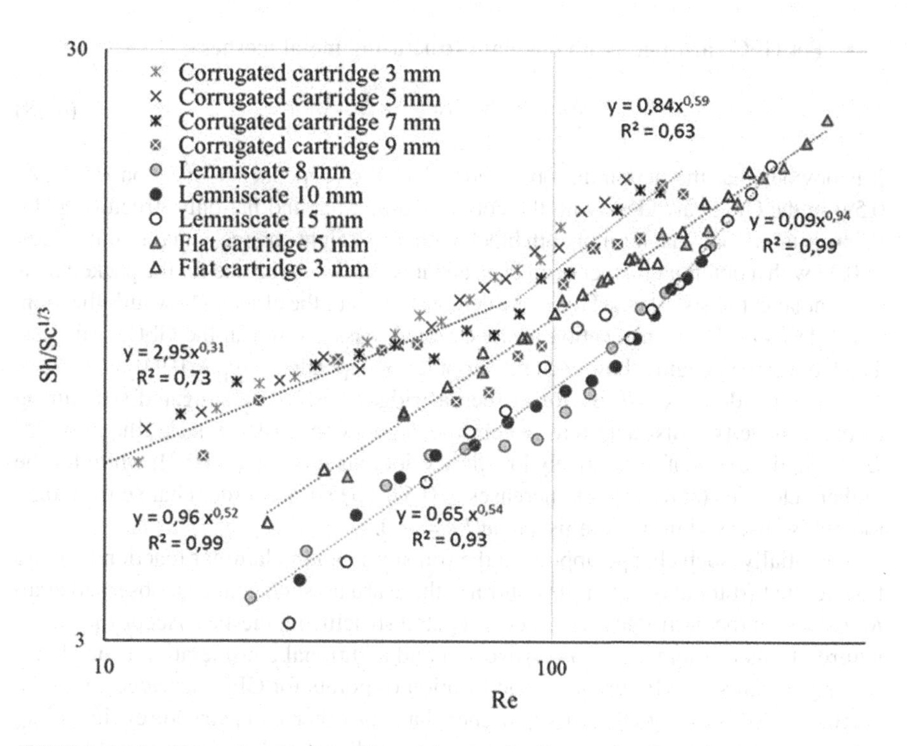


**FIGURE 4.33** Dependence of the *Sh* number on the *Re* number for the different types of GFC cartridges.

*Source*: Reprinted with permission from Zagoruiko et al. 2017.[4]

value of $d_{eq}$ is only the specific diffusion distance between the catalyst layers. The results for both types of $d_{eq}$ are given in Table 4.1.

The dependence of the $Sh/Sc^{1/3}$ on Reynolds number for all tested GFC systems is presented in Figure 4.33.

The data can be described by the power order function in the given coordinates. In general, we may recommend the following criteria equations for the calculation of the mass transfer coefficients, made on the basis of the presented approximations:[4]

- For GFC cartridges with the corrugated structuring metal meshes:

$$\begin{aligned} &Re < 80,\ Sh = 2.95 \times Re^{0.31} \times Sc^{1/3}; \\ &Re > 80,\ Sh = 0.84 \times Re^{0.59} \times Sc^{1/3}. \end{aligned} \tag{4.36}$$

- For the lemniscate GFC beds:

$$\begin{aligned} &Re < 160,\ Sh = 0.65 \times Re^{0.54} \times Sc^{1/3}; \\ &Re > 160,\ Sh = 0.09 \times Re^{0.94} \times Sc^{1/3}. \end{aligned} \tag{4.37}$$

- For GFC cartridges with the plain structuring metal meshes:

$$Sh = 0.96 \times Re^{0.52} \times Sc^{1/3}. \tag{4.38}$$

It is obvious that the maximum observed order ($n$) of dependence of $Sh$ on $Re$ (0.52–0.59) in the GFC cartridges with the corrugated-and-flat and flat-only structuring elements is more than that for monolith block with the straight channels (maximum value[26] of 0.45 with a potential further decrease), and it is similar to the one for the packed beds of granular catalysts[8] (0.64). This is the demonstration of the efficient flow turbulization, which leads to the intensification of the external mass transfer in the GFC cartridges. The lemniscate systems show even higher order of dependence (up to 0.94).

The dependence of $Sh$ on $Re$ in the cartridges with the corrugated structuring elements have two distinct areas, where the dependence order $n$ is significantly different: in the areas of respectively low flow velocities it is equal to 0.31, while for the higher velocities (at $Re > 80$) it increases to 0.59. The similar order change from 0.54 to 0.94 is observed in the lemniscate at $Re = 160$.

Potentially, such change applies to the transfer from the laminar reaction mixture flow to the turbulent one. On the contrary, there are no such changes, observed at all $Re$ values, in the cartridges without corrugated structuring meshes. Accordingly, the nature of this phenomenon is controversial and additional consideration is needed.

As you can see in Figure 4.34, the location of points for GFC cartridges with the corrugated structuring elements is higher than for other GFC structures, including lemniscates. At first appearance, it contradicts the beneficial performance of lemniscate GFCs. This discrepancy, in fact, is only visual and caused by the specific manner of $Re$ number calculation for cartridges with the corrugated meshes.

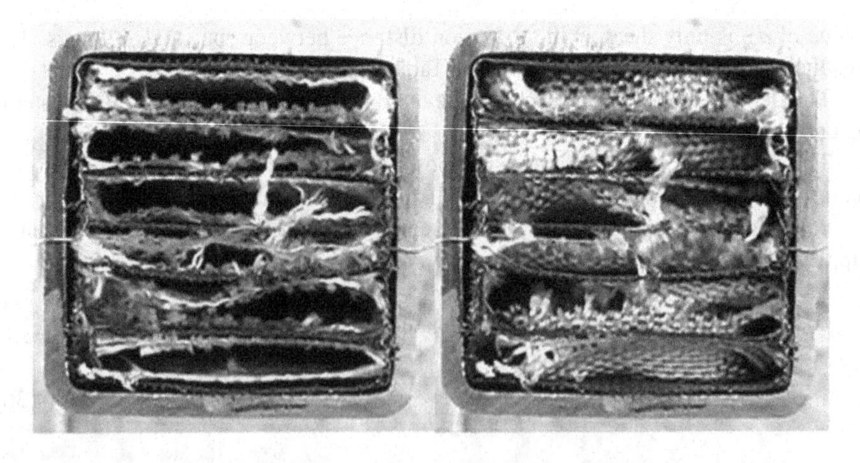

**FIGURE 4.34** GFC fabric geometry change in the structured cartridges under the gas flow wind loading. The initial state is on the left; on the right, under the high-speed flow. Superficial air velocity ~10 m/sec, ambient temperature and pressure.

*Source*: Reprinted with permission from Zagoruiko et al. 2017.[4]

Using the likeness of heat and mass transfer,[8] the obtained criteria equations with the stipulated parameters can be also changed to the equations for the calculation of the heat transfer coefficients in the structured GFC cartridges.

The high apparent activity of the GFC structures is subject to a further consideration. For example, the experimental efficiency of GFCs is considerably higher than for the wire mesh monolith. It seems extraordinary, as the internal geometry of the GFC cartridges with the corrugated meshes is very similar to the wire mesh cartridge, having the same level of flow turbulization. Besides, all external surface of the wire mesh block elements is catalytically active, while in the GFC cartridge the considerable fraction of the surface occupied by the structuring meshes, which don't have catalytic activity at all; therefore, theoretically we can predict higher activity in case of wire mesh monolith.

In our view, this explanation is directly related to the main evident physical distinction of GFCs from all conventional solid catalytic systems, it is due to their mechanical flexibility and mobility. GFC threads and fabrics may interact with the reaction fluid and change their shape, with a potential influence on mass transfer efficiency.

This hypothesis was approved by the experimental monitoring of the GFC fabric behavior under the influence of the gas fluid; the prominent distortion of the fabric shape is clearly seen at high gas flow velocity (Figure 4.34).

Another experiment was performed to study the fabric behavior under moderate gas velocities, general for many practical applications. The movement of the catalyst textiles was not clearly mentioned under such conditions. For that reason another method was applied. The fluorescent paint was sprayed in the air flow at the inlet of the GFC cartridge with the corrugated structuring meshes during the experiment. Figure 4.35 shows the obtained diffusion of the paint particles, observed in UV rays.

The external fabric surface, directed to the gas passage and corrugated meshes side, have a clearly visible paint trace, surrounded along the channels between the corrugated elements. Oddly enough, the internal fabric surface, directed to the flat

**FIGURE 4.35** View of the GFC fabric in the UV rays: external (on the left) and internal (on the right) fabric sides.

*Source*: Reprinted with permission from Zagoruiko et al. 2017.[4]

structuring mesh, also have a rather large amount of paint particles. The paint particles are large enough, so there is no possibility of their diffusion through the fabric. Therefore, it seems fair to say that the convective gas flow in the GFC cartridges is present not only in the cartridge channels, but also, unexpectedly, between the GFC fabric and the supporting wire mesh. Such internal convective flow exists due to the flexibility of the GFC fabric and its mechanical interaction with the reaction fluid, resulting in the shape change.

The described mobility makes the specific surface area of the GFC ($S_{sp}$), which is always supposed to be constant in the traditional approach, the variable value, related to the fluid velocity and the catalyst cartridge geometry. This phenomenon is exclusively attributed to the mechanical flexibility of GFCs, and it is impossible in case of the solid conventional catalyst packing.

The observed behavior explains not only the high apparent activity of GFC cartridges, but also the increase of the order $n$ in dependence of $Sh$ on $Re$ with a higher flow velocity, as specified in the previous chapter. A similar phenomena is also observed in the lemniscate GFC structures, where the increase of the effective catalyst surface area connected with the increase of fluid velocity may lead to "fluffing" of the lemniscate threads, causing the significant increase of its external surface area. Such fluffing, again, may define the order change of the dependence between $Sh$ and $Re$ and its abnormally high value at high $Re$ values. In this context, it seems quite logical for the GFC cartridges with the flat structuring meshes only, that there is no change of the $Sh/Re$ dependence order with fluid velocity, as the catalyst is strongly connected with the wire mesh by clips, thereby limiting the GFC mobility.

### 4.3.7. Verification of the Mass Transfer Limitation Model

To verify the proposed approach for the description of the mass transfer phenomena in GFC cartridges, the toluene conversions were calculated to reproduce the experimental data. The Sherwood values were calculated using the equations (4.36)—(4.38), with the following determination of the mass transfer coefficients via transformation of equation (4.30):

$$\beta = \frac{D \times Sh}{d_{экв}}, \left(m/sec\right) \tag{4.39}$$

Then the apparent rate constant $k^{**}$ was determined using the equation (4.26) followed by the calculation of conversion $X$:

$$X = \left(1 - e^{-k^{**}\tau}\right) \tag{4.40}$$

where $\tau$ is the residence time (sec).

Comparisons of the calculated and experimental data for various GFC cartridges are presented in Figures 4.36–4.38.

A high congruence between the calculated data and experimentally measured ones is observed for all GFC structures. The medium value of the relative error of

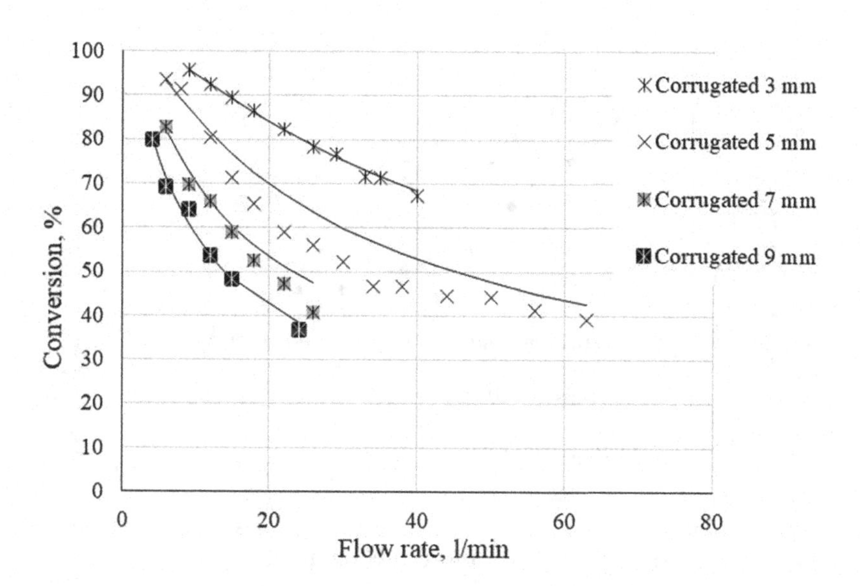


**FIGURE 4.36** Experimental (points) and calculated (lines) values of the toluene conversion upon the gas flow rate at 350°C in the GFC cartridges with the corrugated and flat structuring meshes.

*Source*: Reprinted with permission from Zagoruiko et al. 2017.[4]

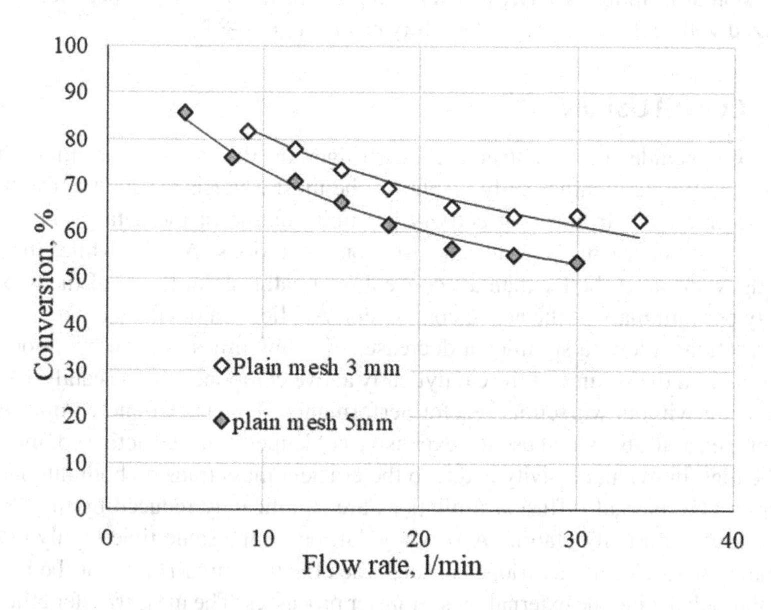


**FIGURE 4.37** Experimental (points) and calculated (lines) values of the toluene conversion upon the gas flow rate at 350°C in the GFC cartridges with the flat structuring meshes only.

*Source*: Reprinted with permission from Zagoruiko et al. 2017.[4]

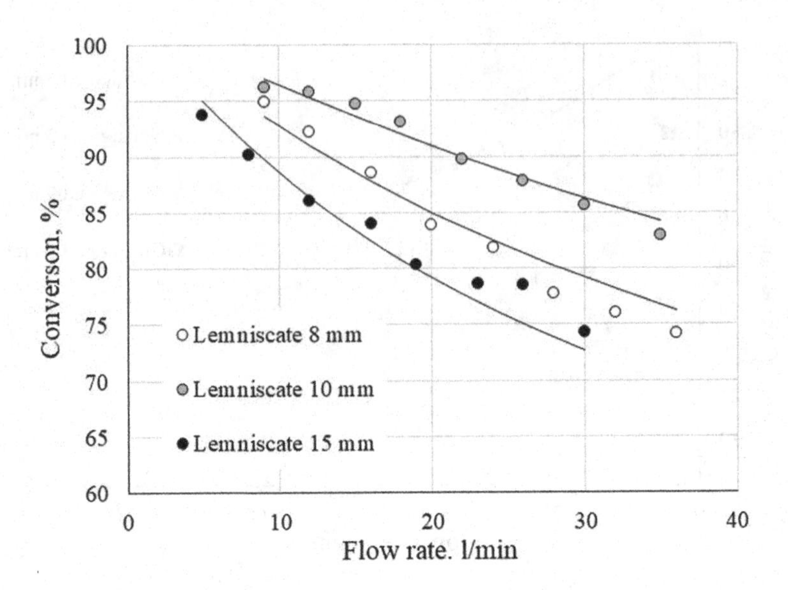


**FIGURE 4.38** Experimental (points) and calculated (lines) values of the toluene conversion upon the gas flow rate at 350°C in the lemniscate GFC structures.

*Source*: Reprinted with permission from Zagoruiko et al. 2017.[4]

conversion description is 7.4%, which is a high accuracy for such experiments, characterized with rather big irreproducibility error (up to 5–8%).

## 4.4. CONCLUSION

We may conclude that the structured cartridges on the base of the microfibrous catalysts are quite commercially viable for the most efficient types of conventional catalysts due to their apparent activity per unit volume of the catalyst bed under the strong influence of the external diffusion limitations. All the while, the GFC cartridges are much better than all conventional catalysts in terms of the apparent activity per unit mass of the active component. Application of GFCs in the fast catalytic reactions allows a significant decrease (by a few times or even by an order of magnitude) in the volume of the catalytically active component in the catalyst loaded into reactor without worsening reactor performance. This is extremely important in case of the catalysts, which use the expensive noble metals as the active components.

The high individual activity is due to the efficient mass transfer, both internal and external. The internal diffusion limitations are significantly reduced by the application of a very thin GFC fabric. A very regular, but at the same time highly uneven, internal structure of the cartridges ensures the effective turbulization of the reaction flow, thus advancing the external mass transfer processes. The mass transfer efficiency is also improved by the change of the GFC fabric shape under the influence of moving fluid, due to the unique flexibility and mobility of GFCs. In turn, this may result in an increase of the effective catalyst external surface in case of a flow velocity increase.

Another important preference of GFC cartridges is their extremely low pressure drop, which is due to their regular inner geometry and high porosity. All types of GFC packing exceed the traditional catalytic systems in a practically valuable ratio between the apparent reaction rate and specific pressure drop.

In general, it is possible to claim that the structured GFC cartridges seem to be one of the most effective catalyst shapes (or even the most effective) among all known types of catalytic structures in the combination of the practically important characteristics (mass transfer intensity, use of the potential of the active components, specific pressure drop).

The lemniscate-shaped structures demonstrate the most advantageous properties amongst all types of GFC packing. The structured GFC cartridges with the flat and corrugated metal meshes and plain GFC fabric are quite competitive as well, though they are slightly less efficient. Notably, the multilayered GFC stack with the propagative flow, which is the most conventional type of GFC packing, characterized with a lower mass transfer efficiency and high pressure drop, shows the worst performance among all GFCs.

The major advantages of the structured GFC cartridges are clearly visible in case of fast catalytic reactions, which occur under a strong influence of the diffusion limitations. Such cases engage the oxidation reactions of CO and volatile organic substances in the gas phase, as well as various reactions in the liquid or multiphase medium. Another potentially interesting area for GFCs application is the complex reaction systems, where the selectivity and yield of the target products are responsive to the mass diffusion limitations.

## REFERENCES


1. Chub O.V. 2009. Investigation of mass transfer processes in glass-fiber catalytic systems. PhD Thesis. Novosibirsk: Boreskov Institute of Catalysis.
2. Chub O.V., Borisova E.S., Klenov O.P. et al. Research of mass-transfer in fibrous sorption-active materials. *Catal Today*. 105: 680–8.
3. Bogoslovskii V.N. (Ed.) 1976. *Heating and ventilation: Textbook for high school.* Moscow: Stroizdat.
4. Zagoruiko A.N., Lopatin S.A., Mikenin P.E. et al. 2017. Novel structured catalytic systems—cartridges on the base of fibrous catalysts. *Chem Eng Proc: Proc Int*. 122: 460–72.
5. Lopatin S.A., Zagoruiko A.N. 2012. Pressure drop of structured cartridges with fiberglass catalysts. Proceedings of the XX International Conference on Chemical Reactors CHEMREACTOR-20 (Luxemburg, December 3–7): 168–9.
6. *Chemist Handbook*. Vol.5. 1968. Moscow/Leningrad: Chemistry.
7. Bensaid S., Marchisio D.L., Fino D. et al. 2009. Modelling of diesel particulate filtration in wall-flow traps. *Chem Eng J*. 154: 211–18.
8. Aerov M.E., Todes O.M., Narinskii D.A. 1979. *Apparatuses with stationary granular bed: Hydraulic and heat basis of operation*. Leningrad: Chemistry.
9. Lopatin S., Mikenin P., Pisarev D. et al. 2014. Pressure drop and mass transfer in the structured cartridges with fiber-glass catalyst. Proceedings of the XXI International Conference on Chemical Reactors "CHEMREACTOR-21" (Delft, The Netherlands, September 22–25): 363–4.
10. Lopatin S., Mikenin P., Pisarev D. et al. 2015. Pressure drop and mass transfer in the structured cartridges with fiber-glass catalyst. *Chem Eng J*. 282: 58–65.

11. Lopatin S.A., Zagoruiko A.N. 2014. Pressure drop of structured cartridges with fiberglass catalysts. *Chem Eng J.* 238: 31–6.
12. Zagoruiko A.N., Glotov V.D., Lopatin S.A. et al. 2016. Investigation of the internal structure, fluid flow dynamics and mass transfer in the multi-layered packing of glass-fiber catalyst in the pilot reactor for sulfur dioxide oxidation. *Science Bulletin of the Novosibirsk State Technical University.* 3(64): 161–77.
13. Ordóñez S., Bello L., Sastre H. et al. 2002. Kinetics of the deep oxidation of benzene, toluene, n-hexane and their binary mixtures over a platinum on $\gamma$-alumina catalyst. *Appl Catal B-Environ.* 38: 139–49.
14. Grbic B., Radic N., Terlecki-Baricevic A. 2004. Kinetics of deep oxidation of n-hexane and toluene over $Pt/Al_2O_3$ catalysts: Oxidation of mixture. *Appl Catal B-Environ.* 50: 161–6.
15. Zagoruiko A.N., Veniaminov S.A., Veniaminova I.N. et al. 2007. Kinetic instabilities and intra-thread diffusion limitations in CO oxidation reaction at Pt/fiber-glass catalysts. *Chem Eng J.* 134: 111–16.
16. Malinovskaya O.A., Beskov V.S., Slinko M.G. 1975. *Modelling of catalytic processes in porous pellets.* Novosibirsk: Nauka.
17. Bal'zhinimaev B.S., Paukshtis E.A., Vanag S.V. et al. 2010. Glass-fiber catalysts: Novel oxidation catalysts, catalytic technologies for environmental protection. *Catal Today.* 151(1–2): 195–9.
18. Bal'zhinimaev B.S., Kovalyov E.V., Kaichev V.V. et al. 2017. Catalytic abatement of VOC over novel Pt fiberglass catalysts. *Top Catal.* 60: 73–82.
19. Bal'zhinimaev B.S., Kovalyov E.V., Kaichev V.V. et al. 2017. Supplementary material for: Catalytic abatement of VOC over novel Pt fiberglass catalysts. *Top Catal.* 60: 73–82.
20. Gulyaeva Y.K., Kaichev V.V., Zaikovskii V.I. et al. 2015. Selective hydrogenation of acetylene over novel Pd/fiberglass catalysts. *Catal Today.* 245: 139–46.
21. Angell C.A. 1988. Perspective on the glass transition. *J Phys Chem Solid.* 49: 863–71.
22. Zanotto E.D., Gupta P.K. 1999. Do cathedral glasses flow? Additional remarks. *Am J Phys.* 67: 260–2.
23. Stokes Y.M. 2000. Flowing windowpanes: A comparison of Newtonian and Maxwell fluid models. *P R Soc A.* 456: 1861–4.
24. Vannoni M., Sordini A., Molesini G. 2011. Relaxation time and viscosity of fused silica glass at room temperature. *Eur Phys J E.* 34: 92–6.
25. Zagoruiko A. 2018. On the intra-fiber mass transfer limitations in glass-fiber catalysts. *Chem Eng J.* 346: 34–7.
26. Reichelt E., Heddrich M.P., Jahn M. et al. 2014. Fiber based structured materials for catalytic applications. *Appl Catal A-Gen.* 476: 78–90.

# 5 Development and Application of Commercial and Pilot-Scale GFC-Based Processes

## 5.1. GFC-BASED PROCESSES FOR THE ABATEMENT OF TOXIC ORGANIC COMPOUNDS IN WASTE GASES

Prevention of technogenic pollution by hazardous volatile organic compounds (VOCs), such as hydrocarbons, organic acids, spirits, aldehydes, etc., is an important challenge in the protection of atmospheric air. Many of these substances are toxic and represent a significant hazard for humans and the biosphere. Besides, the unpleasant smell caused by pollution significantly decreases the quality of life in the surrounding area.

VOC sources are widespread and can be met in many branches of industry, including oil and natural gas processing, petrochemistry, machinery, agriculture and food industry, pharmaceutics, printing, catering, and many others; this stresses a very high relevance of efficient technologies in this area. The important requirements to such technologies include high environmental, energy, and cost-efficiency. This set of requirements corresponds to the technologies based on deep catalytic oxidation of VOCs into harmless products (carbon dioxide and water) most of all.

The high efficiency of GFCs in the reactions of deep oxidation of CO and organic substances makes them attractive catalytic system for the purification of waste gases from different manufacturing facilities.

Here are some examples of the successful application of GFCs in the processes of VOCs deep oxidation in waste gases of commercial plants and purification of diesel exhausts.

### 5.1.1. Process for VOC Deep Oxidation in the Waste Gases of a Synthetic Rubber Plant

The GFC-based process for the catalytic incineration of waste gases was commercialized at Nizhnekamskneftekhim Co. (Nizhnekamsk, Russia)[1,2] facilities for manufacturing of synthetic rubber and isoprene. The characteristics of the waste gases to be purified are presented in Table 5.1.

**TABLE 5.1**
**Characteristics of the Waste Gases of Synthetic Rubber Manufacturing at Nizhnekamskneftekhim Co. Facilities**

| Parameter | Value |
|---|---|
| Flow rate of waste gases | Up to 15,000 st.m$^3$/h |
| Gas temperature at the reactor inlet | 200–350°C |
| Waste gas pressure at the reactor inlet | ~0.04 barg |
| Waste gas composition: | |
| • Isoprene | 100–1200 mg/m$^3$ |
| • Isobutylene | 70–1400 mg/m$^3$ |
| • Formaldehyde | 1–36 mg/m$^3$ |
| • Dimethyldioxane | 0–8 mg/m$^3$ |
| • CO | 50–8000 mg/m$^3$ |
| • Nitrogen oxides | Less than 5 mg/m$^3$ |
| • Oxygen | Less than 5–6 % vol. |
| • Steam | 40–80 % vol. |
| • Nitrogen, $CO_2$ | Balance |

The process was performed in the existing multishelf commercial incineration reactor R90 (Figure 5.1), characterized by a rather large size (diameter of 4500 mm). Prior to the GFC charging, the reactor was operated using the commercial granular catalyst, composed of copper chromite at alumina support; it was loaded to into the second and third shelves in the total amount of 10 tons (position 5 in Figure 5.1).

The special attention should be paid to the low concentration of oxygen and high content of water vapors in waste gases. In addition, these gases are contaminated with the dispersed solid particles of phosphoric acid, which are used upstream as a catalyst. All these factors negatively affected the efficiency of the conventional catalyst: the average purification degree with $CuCrOx/Al_2O_3$ was not exceeding 70–80%, and the catalyst lifetime was not exceeding one year. Under the conditions of low $O_2$ content and high humidity, the outlet gases contained a significant amount of CO, possibly due to prevailing of steam conversion reactions over the deep oxidation.

To solve this problem, it was decided to additionally charge the GFC IC-12-S102, containing platinum in an extra-low amount (~ 0.02 % mass), supported at Zr-modified glass-fiber cloth 850 mm in width, instead of reloading of the conventional catalyst.

Due to some technical reasons, it was decided to load the GFC into the lower shelf in the reactor (position 5 in Figure 5.1). The charging of the radial bed was impossible in this case. Though multilayered packing with the propagative flow was technically possible, it was not optimal due to the possibility of formation of flow non-uniformities across the reactor sequence. Moreover, such loading was undesirable because of the risk of catalyst clogging by the solid particulates and dust. Therefore, the GFC was manufactured and charged in the form of structured spiral cartridges, consisting of GFC cloth and structuring metal elements, to provide the

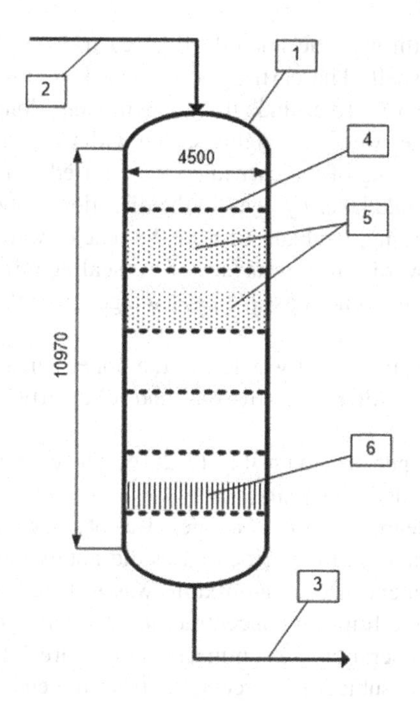

**FIGURE 5.1** Reactor R90 at Nizhnekamskneftekhim Co. facility.

*Source:* Reprinted with permission from Zagoruiko at al. 2010.[1]

convenient loading and prevent the spatial flow non-uniformities. Two types of the cartridges were considered: with corrugated meshes and with volumetric chain-link metal net structures (see Figure 3.7).

The attempts to produce the cartridge with a height equal to the standard width of the glass-fiber fabric (~850 mm) faced difficulties in producing the corrugated mesh of such width, as it was possible to produce the corrugated bends without the deformation of mesh structure, only not wider than 200–250 mm. Manufacturing of the cartridges with such height brought the need to cut the glass-fiber fabric, which was not desirable, as it resulted in a significant decrease of the mechanical stability of the glass-fiber support. Therefore, the optimal option was the one based on chain-link metal net.

However, it was not so easy to produce the spiral cartridge with the external diameter equal to the internal reactor diameter (4.5 m) and, moreover, it would be practically impossible to load such a cartridge into the reactor due to the necessity to load it through the lids with 600 mm diameter. Therefore, it was the first case when the modular principle of GFC bed formation from cartridges of relatively small size and convenient in manufacturing, transportation, and handling was applied in commercial practice. The external diameter of such cartridge was equal to 380–420 mm (to ensure loading through the lid) with a height of ~900 mm. The GFC amount in each cartridge was equal to ~12 kg; 81 cartridges were finally loaded into R90 reactor, which corresponds to the overall GFC mass of ~1 ton.

Cartridges were loaded into the reactor through the side lids. Each cartridge was positioned vertically at the lower reactor shelf. The cartridges were packed close and pressed down to each other (see Figure 3.8). To exclude the gas bypassing along the catalyst, all spaces between the cartridges were thoroughly sealed with the pads made of thermostable mineral wool. Positioning of the cartridges was started from the reactor shelf center and then they were laid layer-by-layer in the direction of the reactor walls. After the completion of charging, void spaces near the reactor walls were also sealed with mineral wool. Afterwards, all top surface of the sealing mineral pads was flattened with putty to avoid any other possible gas leakages from the GFC bed.

The preceding loading of the conventional catalyst was left in the upper part of reactor; the beds of these catalysts served as a filter to remove dust and solid particulates from the gas flow.

The first start-up of the R90 reactor was performed on July 17, 2008. The overall time onstream for the first GFC batch exceeded four years.

During the whole operating period the degree of VOC abatement was observed at the level of 99.5–99.9%; residual VOC content in the purified gases did not exceed 10–15 mg/m$^3$. In all cases the residual content of each admixture was well below both the industry and state standards for the limiting concentrations of hazardous substances in air. The results of four years operation are summarized in Figure 5.2.

It was also very important to receive the subjective perceptions from the enterprise employees about the absence of typical odors at the plant site, which were present during the whole period of the production on the base of conventional $CuCrO_x/Al_2O_3$ catalyst only.

During more than four years of operation, there was no decrease of the purification degree, which reaffirmed both the high efficiency of IC-12-S102 GFC and its high resistance to the deactivation in dusty media.

It is necessary to note that the IC-12-S102 GFC bed occupies less than 10% of the reactor R90 volume. This reactor was initially designed for the application of

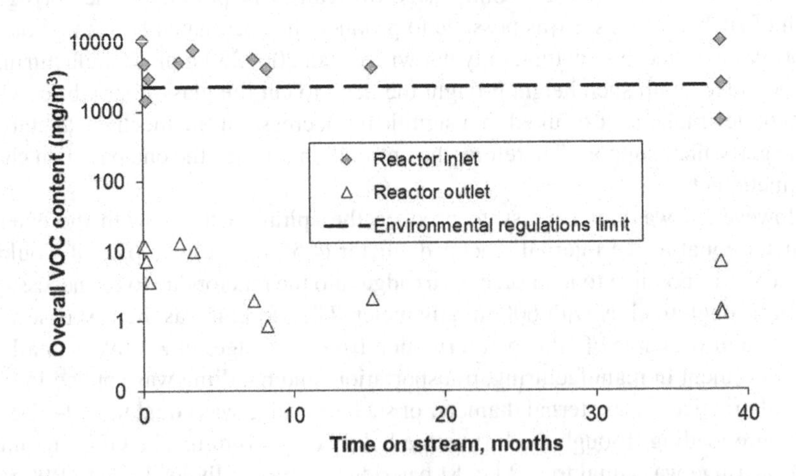


**FIGURE 5.2** GFC-based process performance during long-term operation.

the conventional granular catalysts and was not optimal for GFC. It provides the significant potential to decrease the reactor size and metal consumption in case of purposeful optimization of the reactor design for GFCs application. In general, it was concluded that the GFC-based process persuasively demonstrated its efficiency and superiority over the conventional technology.

### 5.1.2 Process for the Purification and Cooling of the Exhausts from a Stationary Diesel Power Plant

The ability of Pt-containing GFCs to oxidize carbon monoxide and hydrocarbons efficiently, as well as to reduce nitrogen oxides, makes them appropriate catalysts for the purification of gaseous exhausts from internal combustion engines.[3,4]

Pt-containing GFC IC-12-S111, described in Chapter 2, was used to create a system for the purification and cooling of exhaust gases from a 630 kW stationary diesel electricity generator MTU12V2000G63 involved in the emergency energy supply in the Moscow subway.

The exhaust gases of this engine have an outlet temperature of 450–590°C and a gas flow rate of 0.70–0.85 st.m³/sec; excessive pressure is not higher than 10 kPa. These gases contain the following hazardous admixtures (in mg/m³): nitrogen oxides ($NO_x$)—1400–2000; carbon monoxide (CO)—150–650; hydrocarbons ($C_xH_y$)—30–150; solid particulates (soot)—10–50.

In addition to the requirements regarding the quality of gas purification, established in accordance with the state standards on diesel emissions (GOST 17.2.2.02–98 and GOST R 41.96–2005), it was necessary to cool the purified gases to a temperature not more than 120°C.

The flowsheet of the proposed process, including the catalytic neutralizer, vortex scrubber, and drip trap, is shown in Figure 5.3.

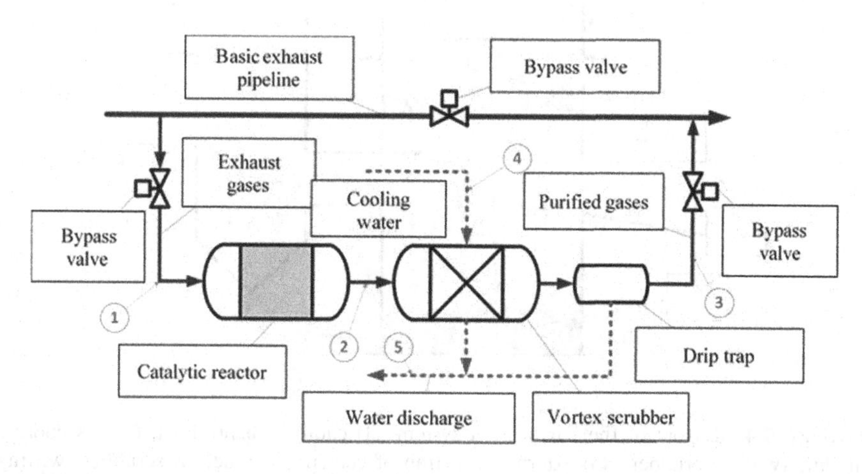


**FIGURE 5.3** Flowsheet of the GFC-based process for the purification and cooling of exhaust gases. 1–5 are points of process parameters control.

The exhaust gases from the diesel engine with the temperature of 450–590°C pass through the catalytic reactor, where CO and hydrocarbons are oxidized into carbon dioxide and water:

$$CO + O_2 \Rightarrow CO_2$$

$$C_xH_y\,(\text{hydrocarbons, soot}) + O_2 \Rightarrow CO_2 + H_2O$$

Also, the reaction of $NO_x$ reduction by the components of the reaction mixture may occur here:

$$CO/C_xH_y + NO_x \Rightarrow N_2 + CO_2 + H_2O$$

Then the exhaust gases intensively contact with the water in the vortex scrubber. It results in the decrease of the gas temperature to a temperature not exceeding 115°C; in addition, water absorbs some part of the nitrogen and sulfur oxides, as well as soot particulates, thus providing the additional gas cleanup.

After that the gaseous flow passes through the drip trap to separate the drip water, and the spent water is poured out of the scrubber and drip trap into the drainage pipeline. The flow of the purified and cooled exhaust gas is discharged into the atmosphere.

The system is installed in bypass to the basic exhaust line.

To decrease the size and metal consumption in the purification system, it was decided to combine the catalytic neutralizer and vortex scrubber into one reactor shell, with the separate placement of the drip trap (Figure 5.4).

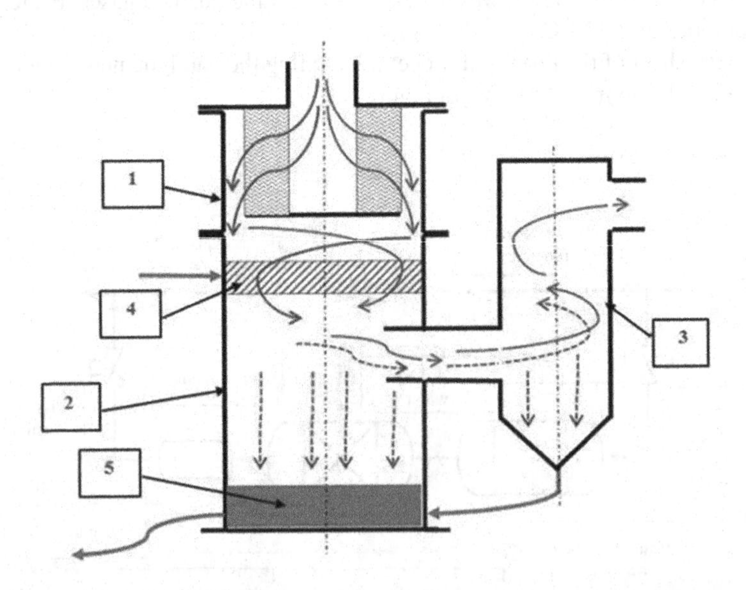


**FIGURE 5.4** Layout of the purification system: (1) catalytic neutralizer, (2) gas cooling module (vortex scrubber), (3) external drip trap of centrifugal type, (4) scrubber swirling device, (5) wastewater collection tank. Gas flow is shown by solid lines; drip water fluids by dashed ones.

The external appearance of CS-3000 and its internal layout are presented in Figure 5.5.

As shown in Figure 5.5, the system consists of cylindrical shell (1) with an internal diameter of 1000 mm, the cassette with 24 GFC cartridges (4), and the gas distribution system with the external perforated plate (5) and distributing plates (6). Admission of the exhaust gases is performed via inlet pipeline 10 (internal diameter 426 mm). The centrifugal vortex scrubber consists of the cylindrical shell (2), the swirling device (7), and the internal drip trap (8). The shell bottom (9) has a conical shape to simplify the water discharge and prevent its accumulation in the bottom part of the shell. The gas flows from GFC cassette to the swirling device (7), where it is contacted with the water, supplied via the fitting (13) (two 1.5″ fittings located on opposite sides of the swirling device). Drip water separated in the internal drip trap (8) is poured out to the drainage pipeline via the fitting (15). Gas-liquid flow from the scrubber enters the external drip trap (3) via the connecting pipeline (12). Water-free gases are passed into the atmospheric discharge pipeline via the outlet pipeline (11) (internal diameter is 426 mm); separated water is poured out via the branch pipe (14) back to the scrubber shell following discharge to drainage.

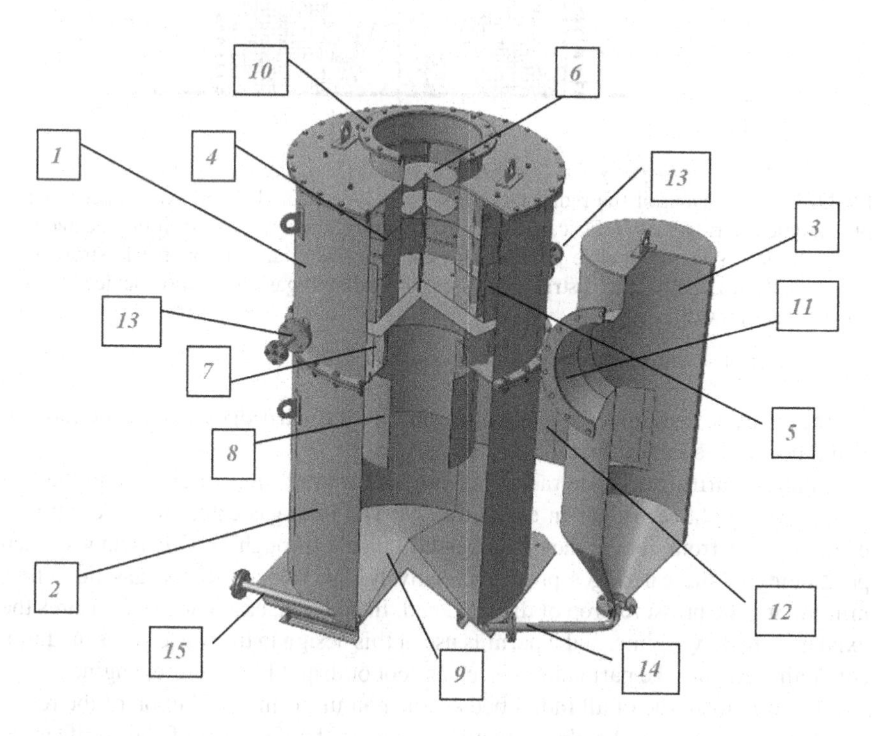


**FIGURE 5.5** External appearance of the purification system with its internal layout: (1, 2) cylindrical shells, (3) external drip trap, (4) catalytic neutralizer, (5) perforated ring, (6) distribution plates, (7) swirling device, (8) internal drip trap, (9) conical bottom, (10, 11) fittings for the supply and removal of exhaust gases, (12) connecting pipeline, (13) fittings for water supply, (14, 15) pipeline and fitting for water discharge.

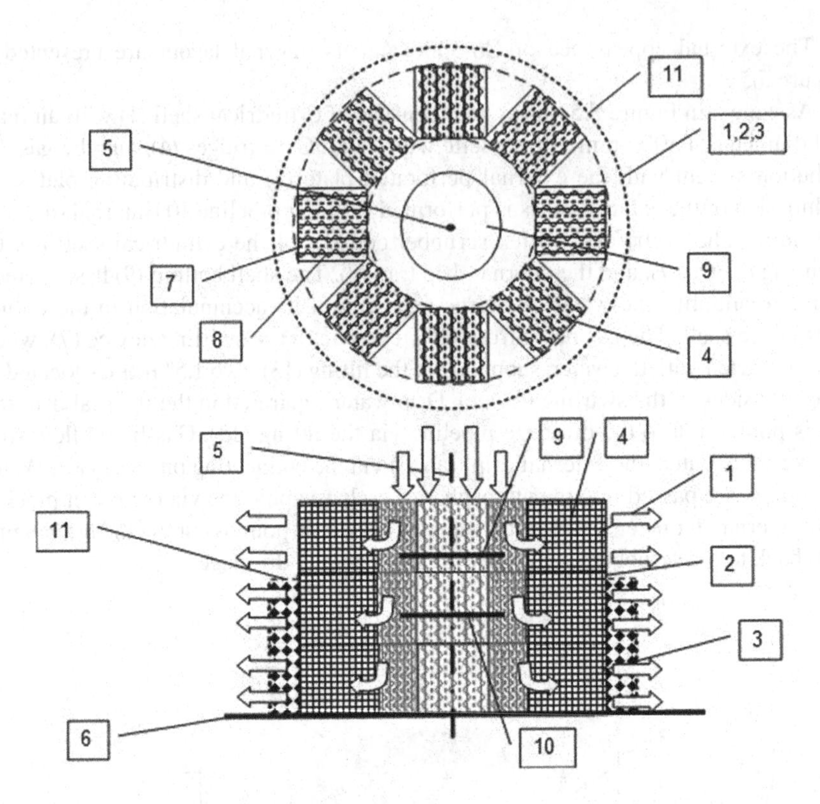


**FIGURE 5.6** Layout of the radial GFC-based catalytic bed: (1, 2, 3) axisymmetric rings of catalytic cartridges, (4) GFC cartridges, (5) system symmetry axis, (6) impermeable partition, (7) layers of the flexible microfibrous catalytic material, (8) volumetric structuring elements, (9, 10) internal gas distribution discs, (11) external gas distribution device. Arrows show the gas flow directions.

*Source:* Reprinted with permission from Lopatin et al. 2015.[9]

The catalytic neutralizer is made of prismatic GFC cartridges, assembled into the radial bed[5–8] (see Figure 5.6).

Catalytic cartridges in the radial bed are arranged in three circles along the bed height, with eight cartridges in each circle.[9,10] The inlet gas enters the internal part of the reactor from above, then it is passed outside through GFC cartridges. Such positioning of the cartridges provides relatively low velocity of the gas flow, thus minimizing the pressure drop of the GFC bed. In addition, each cartridge ensures the axial gliding flow regime, and it permits use of this design in the waste gases medium, contaminated with the particulates, such as soot or drip oil from diesel engine.

The common issue of all radial beds is the non-uniform distribution of the reacting flow along the bed height, potentially causing the decrease of the purification efficiency. This problem was solved in the presented reactor by the design optimization and development of the additional distributing devices on the base of computational fluid dynamics modeling. It was shown that the optimal design includes the additional gas distributing elements, in particular, the flat distributing disc in the internal part of the bed and external round retaining grid, made of perforated metal tape[6] (see Figure 5.7).

As was shown in Chapter 4.2.3, GFC cartridges are characterized by spatial anisotropy. Due to the significant difference in ratios $\xi_\theta/\xi_a$ and $\xi_\theta/\xi_a$ (see Figure 4.19) the spatial orientation of the cartridge in the neutralizer is important. As long as main channels in the cartridge are oriented horizontally along the flow in all cases, it is necessary to distinguish the orientation of the catalyst planes (see Figure 5.8).

Both neutralizer and scrubber have an axial symmetry and there are no reasons for the generation of gas flow non-uniformities in respect to the internal perimeter of

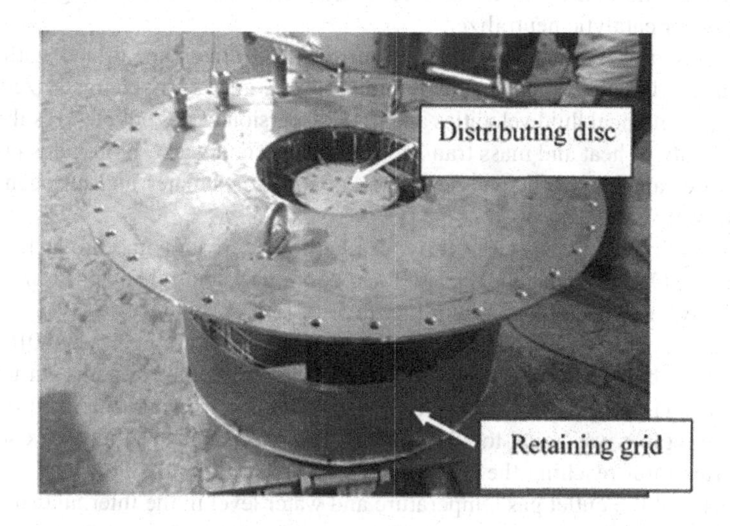


**FIGURE 5.7** Appearance of GFC-based modular radial bed catalytic neutralizer with the gas distributing devices. Arrows show the gas flow directions.

*Source:* Reprinted with permission from Lopatin et al. 2015.[9]

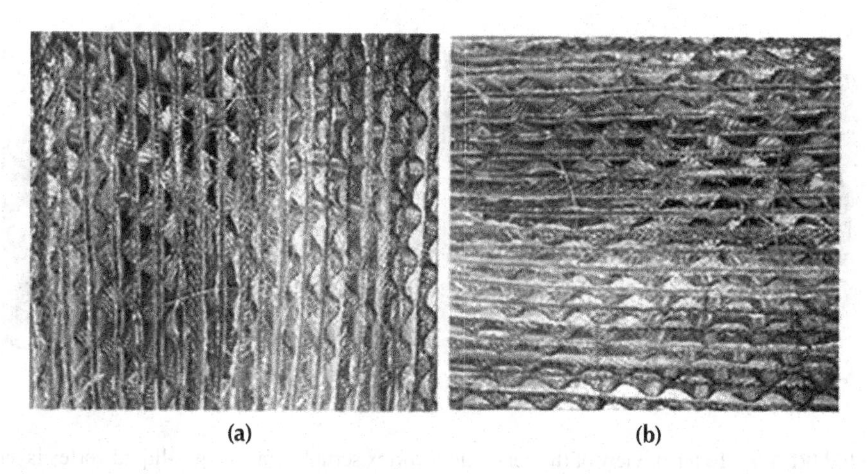


**FIGURE 5.8** Possible orientation of the catalyst planes in prismatic cartridges.

radial bed of GFC cartridges. Therefore, the optimization of the bed design should aim at the uniformity of flow distribution in a vertical direction only. In case of the vertical positioning of the catalyst planes (Figure 5.8a), the hydraulic resistance coefficient in a vertical direction is rather low; this permits redistribution of the flow inside the channels, thus improving the uniformity of the velocity distribution along the neutralizer height. Such redistribution is negligible in case of the horizontal positioning of GFC planes (Figure 5.8b) due to a higher hydraulic resistance perpendicular to the planes, so that no flow leveling occurs. Therefore, the vertical positioning of GFC in the cartridge is optimal.

Figure 5.9 shows the centrifugal vortex scrubber, used for the cooling and cleanup of gases after catalytic neutralizer.

In this scrubber the waste gases from the neutralizer are mixed with water. Their interaction gives rise to a gas-liquid vortex (Figure 5.9a), characterized by the extremely high local fluid velocities and high dispersion of gas bubbles, resulting in a high intensity of heat and mass transfer between gas and water. It provides fast cooling of gases and their efficient cleanup from soot particulates and nitrogen oxides (mostly, well-soluble $NO_2$).

This development has successfully resulted in the creation and production of six commercial CS-3000[10] gas purification and cooling systems. The appearance of the CS-3000 system is demonstrated in Figure 5.10.

The acceptance testing of these systems was performed in February 2016. At the startup the exhaust gases were directly fed into the purification systems without their preheating. The duration of the catalytic neutralizer heating and stabilization of the operating regime was equal to 15 minutes. The water supply to the vortex scrubber was started after reaching the steady-state reaction regime; the flow was controlled on the base of the outlet gas temperature and water level in the internal drip trap, to exclude overflow in the bottom part. Duration of continuous operation in the tests was not less than 30 minutes. The experimental procedure included the continuous monitoring of the gas temperature and composition at the system inlet and outlet, as well as the system pressure drop.

(a) (b)

**FIGURE 5.9** Exterior view of the centrifugal vortex scrubber (a) and gas-liquid vortex layer inside it (b).

**FIGURE 5.10** System CS-3000 for the purification and cooling of the exhaust gases from a stationary diesel generator in commercial operation.

Due to objective reasons, each diesel-generator had not more than 350–370 kW loading (i.e., less than 55–60% of the nominal operation capacity). This, in turn, led to a significantly decreased (approximately twice below the nominal) gas flow rate and low temperatures in the GFC bed (310–350°C instead of the nominal 450–490°C). Such a regime is far from optimal for CS3000 systems, as long as the decreased flow rate has a negative influence on the efficiency of the vortex scrubber, while the decreased temperature leads to a lower GFC activity in the reactions of CO and hydrocarbons oxidation, and especially in the $NO_x$ reduction reaction.

The measured pressure drop of the system was not exceeding 80–90 mm WC. No signs of increased hydraulic resistance, indicating the potential clogging of GFC cartridges and other system elements by soot deposition, were observed during the tests.

The stable cooling of the outlet gases down to 110–150°C was provided by the water supply to the vortex scrubber.

The residual content of CO in the outlet gases did not exceed 43–47 ppm (54–59 mg/m$^3$); CO removal degree was 75–78%. The residual content of hydrocarbons varied from 0 to 11 ppm (14–22 mg/m$^3$); the conversion degree for hydrocarbons was observed in the range from ~50% to 100%.

Notably, the residual concentration of CO and hydrocarbons in the purified gases appeared to be below the values established in the strictest effective international standards for non-road diesel engines (Tier 4–2015, USA and EU Stage V, European Union). $NO_x$ content was as high as 970–990 ppm (2000–2100 mg/m$^3$); $NO_x$ conversion did not exceed 5–10% due to insufficient temperature. Sulfur dioxide was not found in the purified gases. Measurements of soot content were not performed. The results of tests are summarized in Table 5.2.

In general, all tested CS-3000 systems demonstrated stable operation without failures or accidents. Regarding the purification efficiency, the systems met all requirements of the national standard GOST R 41.96–2005 for new engines at limiting air exchange. The CS-3000 systems also provided the efficient cooling down of the purified gases to the required temperature. Pressure drop of the systems was significantly lower than the assigned limit.

Actual flow rate of the exhaust gases and their temperature were well below the nominal values due to the insufficient loading of generators during the tests, and it had a negative influence on the efficiency of gas purification, especially, in respect to $NO_x$ abatement. Hopefully, this efficiency can be increased at the nominal operation regimes.

**TABLE 5.2**
**Summary of the CS-3000 Testing Data**

| Parameter | Requirement | Fact | Agreement with requirements |
|---|---|---|---|
| **Initial conditions** | | | |
| Gas flow rate, st.m$^3$/h | 3000* | 1650 | *Below requirements due to insufficient generator loading* |
| Gas temperature at the system inlet, °C | 450–490* | 310–350 | |
| Content of admixtures (mg/m$^3$): | | | |
| Nitrogen oxides ($NO_x$) | 1400–2000* | 2100–2300 | *Exceeds* |
| Carbon monoxide (CO) | 150–650* | 185–220 | In agreement |
| Hydrocarbons ($C_xH_y$) | 30–150* | 40–60 | In agreement |
| **Test results** | | | |
| System pressure drop mm WC | Below 500** | Below 90 | In agreement |
| Outlet gas temperature, °C | Below 115** | 110 | In agreement |
| Content of admixtures (mg/m$^3$): | | | |
| Nitrogen oxides ($NO_x$) | 9.0*** | 8.5–9.0 | In agreement |
| Carbon monoxide (CO) | 1.5*** | 0.06–0.07 | In agreement |
| Hydrocarbons ($C_xH_y$) | 4.0*** | 0–0.1 | In agreement |

*Remarks:*
* Parameters of MTU12V2000G63 diesel-generator, according to the information from manufacturer.
** Technical assignment for purification/cooling system.
*** According to Russian diesel emission standard GOST R 41.96–2005.

According to the obtained results, all CS-3000 systems were put into commercial operation.

## 5.2. GFC-BASED PROCESSES OF ENVIRONMENTALLY SAFE COMBUSTION OF FUELS

Nowadays, processes for the environmentally-friendly combustion of various fuels is an important area for the application of catalytic technologies. The most relevant tasks in this area include the achievement of complete fuel combustion, the possibility to use low-quality fuels and combustible wastes for energy production, and minimization of contamination of combustion products by hazardous and toxic substances (CO and hydrocarbons, nitrogen oxides, soot).

### 5.2.1. Combustion of Solid Fuels in the Fluidized Beds of the Dispersed Heat Carrier Using Reinforced GFC Cartridges

This work was devoted to the combustion process of dispersed solid fuels in the fluidized beds of the dispersed heat transfer carrier.[11] This process, mentioned earlier (see Chapter 3.3.3), imposed very serious requirements to the mechanical strength of the catalysts due to a high abrasive action of a solid particles fluidized bed; therefore, it was recommended to use the described reinforced GFC cartridges, such as those shown in Figure 3.13.

The view of the concentric reinforced cartridges manufactured for this process is shown in Figure 5.11. Each cartridge included three coaxially nested reinforced GFC round planes. In total, six cartridges were produced and assembled into the catalytic system with the basic firming steel axis step of 60 mm between the cartridges; the total height was equal to 660 mm.

**FIGURE 5.11** View of the concentric reinforced GFC cartridges.

The experiments were performed at the pilot plant in the Boreskov Institute of Catalysis (Novosibirsk, Russia) in the combustion process of dispersed coal in the fluidized bed of the heat transfer agent (sand). The pilot-plant scheme is demonstrated in Figure 5.12. The experimental program also involved comparative tests with the extra-strong ceramometal catalytic monolith,[12] as well as catalytically inert material (ceramic monoliths) instead of the catalyst.

The reactor had an internal diameter of 75 mm. The sand loading was equal to 3.5 liters. Height of the fluidized bed was maintained in the range 800–900 mm.

As shown by the experiments, the proposed GFC cartridges demonstrated stable performance without signs of clogging by the particulates or traces of abrasive stirring. The results are summarized in Table 5.3.

As seen from the Table 5.3, the reinforced GFC cartridges demonstrated a high efficiency of CO and hydrocarbon oxidation, comparable with the efficiency of the best known conventional catalysts. At the same time, GFCs are more scalable and simpler to manufacture. GFC cartridges can be mechanically stable and resistant

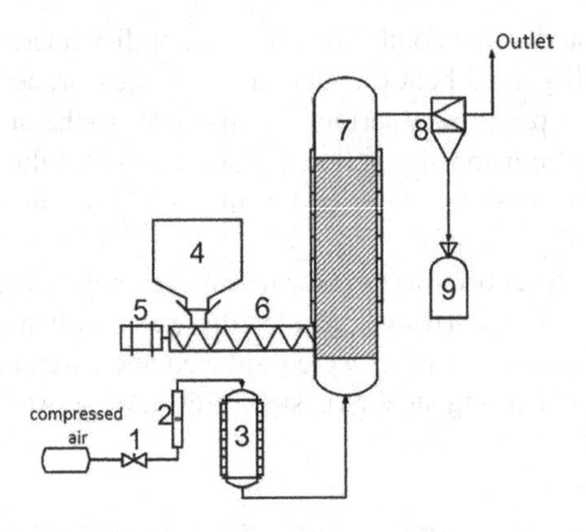


**FIGURE 5.12** The pilot-plant scheme for the combustion of dispersed coal in the fluidized bed of sand particles: (1) valve, (2) rotameter, (3) heater, (4) coal reservoir, (5) electric engine, (6) screw feeder, (7) reactor with fluidized bed and electric heating, (8) cyclone separator, (9) ash collection vessel.

**TABLE 5.3**
**Results of the Experiments with Different Catalysts at the Pilot Coal Combustion Unit at 750°C, Fuel—Coal (particle size less than 1.25 mm)**

| Parameter | Experimental results | | |
|---|---|---|---|
| | Inert monolith | Reinforced GFC cartridges | Ceramometal $CuCrAlO_x$ |
| Residual content of CO in outlet gases, ppmv | 4244 | 584 | 432 |
| Residual content of $NO_x$ in outlet gases, ppmv | 258 | 280 | 503 |
| Residual content of $CH_4$ in outlet gases, % vol. | 0.05 | 0 | 0 |

to stirring, as long as the protective metal mesh is mechanically stronger and has a higher resistance to abrasion than ceramics. In general, GFC-based cartridges look quite promising for the given application and deserve further development.

### 5.2.2. Catalytic Air Heaters on the Base of GFCs

A catalytic air heater was developed for the environmentally-friendly combustion of hydrocarbon fuels. This heater enabled the combustion of two-stage catalytic fuels, combining their flame and catalytic oxidation.[13–15]

The flowsheet of the developed catalytic combustor is presented in Figure 5.13.

Firstly, there is the oxidation of the initial fuel in the burner with a limited air supply to the flame region. A lack of oxygen in this region allows the minimization or complete exclusion of the generation of nitrogen oxides. Concurrently, the conversion of fuel is not enough, which results in the presence of CO in the combustion residues. To overcome this problem the burner flame is mixed with air and sent to a GFC-based cartridge for the catalytic oxidation of CO and unburned by-products to receive harmless products: $CO_2$ and water.

The unit contains the Pt/GFC arranged inside the spiral cartridge, textured with corrugated and plain metal meshes (Figure 5.14). As indicated earlier, the cartridges

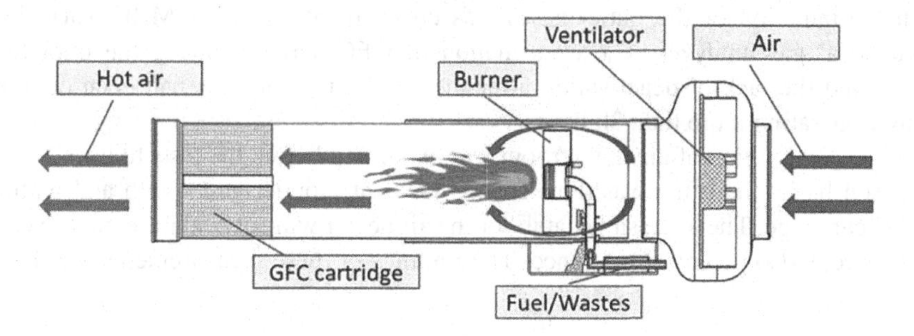


**FIGURE 5.13** The flowsheet of the catalytic combustor.

*Source:* Reprinted with permission from Lopatin et al. 2019.[15]

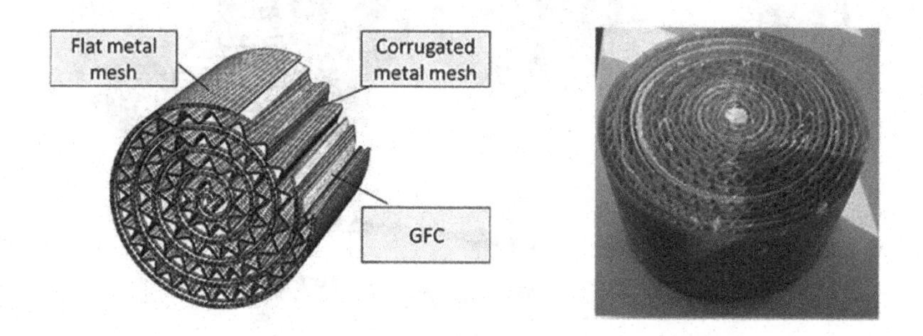


**FIGURE 5.14** The cylindrical spiral GFC cartridge.

*Source:* Reprinted with permission from Lopatin et al. 2019.[15]

have a high potency of mass transfer and low pressure drop.[16] In addition, they have a high resistance to thermal and mechanical shocks, and have respectively high axial and radial heat conductivity, which allows then to distribute temperature in the cartridge evenly during the reaction.

The cartridge, manufactured for the pilot tests, had a corrugation height of 3 mm and an external diameter of 160 mm.

The prototype of the catalytic air heater with the planned heat power of 15 kW (Figure 5.15) was constructed to demonstrate the efficiency of the proposed approach.

Standard automotive LPG ("winter" type with propane/butane ratio 9:1) was used as a fuel.

The pilot tests showed the easy heater startup and stable operation with the following characteristics:

- Actual heat power: ~15 kW
- LPG consumption: 1.2 kg/h
- Air flow rate: 150 $m^3$/h
- Temperature in the catalytic cartridge: 450–550°C
- Temperature of air flow at heater outlet: 250–300°C
- Air flow temperature within 3 m distance from the air heater outlet: Below 80°C

The formulation of the outlet gases was checked by means of MRU Vario Plus Industrial gas analyzer. The examination of GFC activity during the pilot tests approved the lack of deactivation after the combustion process had been continuously operating more than 30 hours.

The constitution of air in the room was examined before the tests to have a comparison basis, the same was done for the air heater outlet gases with and without GFC cartridge. The accessible data[17] for the air heater with a standard monolith catalyst[18] were also taken for reference. The findings of these measurements are shown

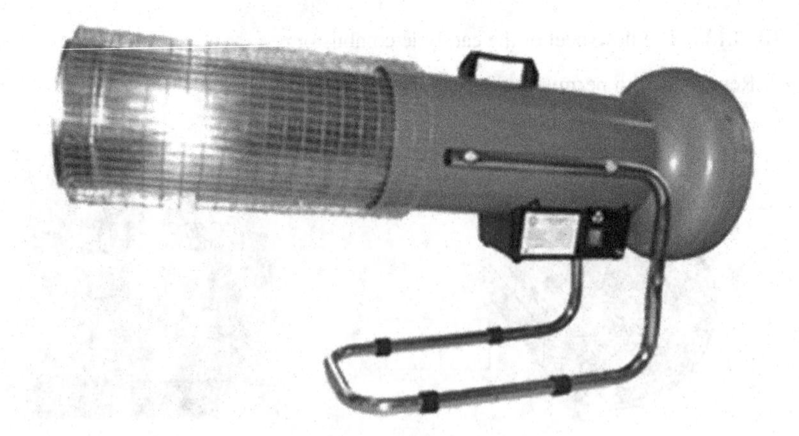

**FIGURE 5.15** Prototype of the catalytic air heater with the heat power 15 kW.

*Source:* Reprinted with permission from Lopatin et al. 2019.[15]

**TABLE 5.4**
**Outlet Gas Composition for the Different Methods of Fuel Combustion**

| Substance | Concentration of substances in outlet gases (mg/m³) | | | | |
|---|---|---|---|---|---|
| | Initial air | Air heater without catalyst | Air heater with conventional monolith catalyst | Air heater with GFC cartridge | Maximum allowable concentration |
| CO | 0 | 10 | 13 | 4 | 20 |
| $NO_x$ | 0 | 23 | 3 | 0 | 5 |
| $C_xH_y$ | 9 | 24 | 30 | 11 | 300 |

*Source:* Reprinted with permission from Lopatin et al. 2019.[15]

in Table 5.4. The figures on residual content of admixtures were also set against a maximum permissible concentration, defined by the legislation for the environmental protection (Russian national standards for the level of air purity in the industrial facilities were deliberated).

It is obvious that the GFC-based air heater ensures the deficiency of nitrogen oxides and minimum occurrence of CO and hydrocarbons in the outlet air. According to organoleptic observations, there was no foreign smell in the room, and you could breathe easily. This means that the air heater ensures the superior environmental efficiency compared to other combustion methods, including those based on conventional catalysts.

The created combustor can be applied for the direct air heating in various premises and facilities. If the initial air comprises any organic compounds, they are also oxidized; thus, the combustor clears the air in the premises. The air can be delivered from the outside, so the combustor unites the characteristics of air heating and ventilation systems.

Apart from the usual gaseous fuels (LPG, natural gas), the GFC combustor can be adjusted for the use of liquid fuels, both the conditional and liquid wastes, having a significant content of toxic organic compounds. In the latter instance, the unit can be applied not only for energy production, but also for the recycling of hazardous wastes. The GFCs are reputedly the efficient catalysts for chlorinated hydrocarbon[19,20] oxidation; for that reason, the application field may engage the recycling of liquid toxic chlororganic wastes without occurrence of the toxic secondary wastes as phosgene or dioxins. Further research should also consider the stability of GFCs and their lifetime under various reaction conditions.

## 5.3. SULFUR DIOXIDE OXIDATION PROCESSES AT PT-CONTAINING GFCs

As shown earlier,[21–27] the platinum-containing GFCs of the previous generation, based on glass-fiber supports promoted with zirconia, demonstrate high activity in oxidation of $SO_2$, especially in the area of low temperatures (< 400°C). As such, they have a high resistance to deactivation in real application conditions. The GFCs of a new generation, described in Chapter 2, have even higher activity and lower temperature of the reaction "ignition."

All of this makes Pt/GFC an attractive catalyst for the development of the novel processes for $SO_2$ oxidation in sulfuric acid production, as well as in the abatement of various flue and waste gases.

### 5.3.1 Processes of $SO_2$ Oxidation in Sulfuric Acid Production

The possibility of using Pt/GFC in existing reactors at traditional sulfuric acid plants was analyzed using the example of $SO_2$ oxidation in a standard adiabatic reactor with four fixed catalyst beds and intermediate cooling of the reaction flow between these beds.[2] This analysis was based on the equilibrium evaluations for adiabatic beds and assumption on the sufficient amount of catalyst to achieve the $SO_2$ oxidation equilibrium in each bed.

Figure 5.15 shows the process trajectories plotted in coordinates "$SO_2$ oxidation degree vs temperature." In the traditional process based on the application of vanadia catalyst, the gas is fed into the first bed at temperature of about 420°C, which is slightly higher than the catalyst ignition temperature (360–400°C, depending on the catalyst type). The oxidation reaction is accompanied with the reaction heat emission, leading to adiabatic temperature increase in this bed (A–B segment of the solid line in Figure 5.16). Then the gas is cooled in the heat exchanger to a definite temperature not lower than 420°C (segment B–C). Afterwards, the gas is successively passing the catalyst beds and heat exchangers, with repeated temperature increase

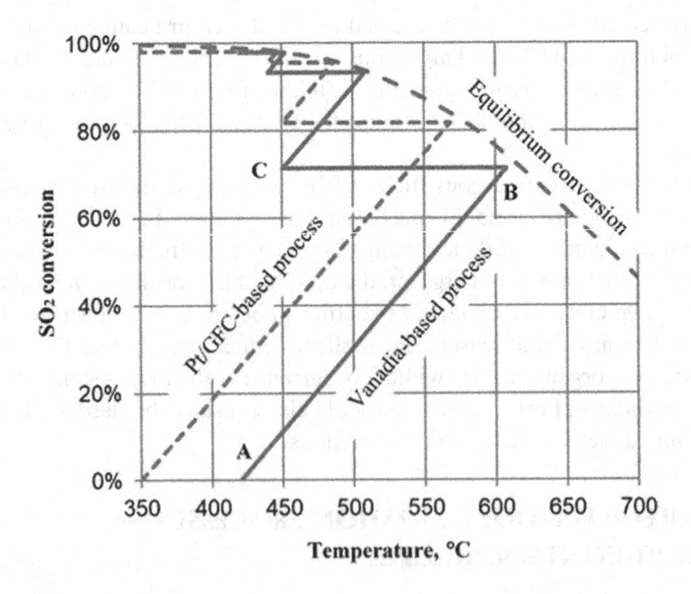


**FIGURE 5.16** Temperature conversion chart for $SO_2$ oxidation process in a four-bed adiabatic reactor with interbed gas cooling. Inlet gas composition: 9% vol. $SO_2$, 14% $O_2$. Solid line shows the trajectory of the conventional process on the base of $V_2O_5$ catalyst; the dashed line, the new GFC-based process.

(sloping segments) and cooling (horizontal segments) at each stage. Under the given conditions the overall sulfur dioxide conversion reaches 98.9%.

Pt/GFC has a low ignition temperature (~300°C), so even with some temperature reserve the inlet gas temperature in the first and last reactor beds can be decreased to 350°C, thus decreasing the outlet temperature and improving the equilibrium conditions (dashed line in Figure 5.16). In this case the overall conversion rises to 99.84%, and it corresponds to the decrease of sulfur dioxide content in waste gases of the plant by more than 6 times.

Figure 5.17 demonstrates the dependence of maximum temperature in the first adiabatic $SO_2$ oxidation reactor on the sulfur dioxide inlet concentration for the different inlet gas temperatures. The high-temperature vanadia catalyst works steadily starting from the inlet temperature of ~420°C, at this the maximum temperature in the bed reaches the thermal stability limit for this catalyst (~650°C) at the inlet concentration of $SO_2$ equal to 13.3% vol. Application of the low-temperature vanadia catalyst allows to decrease the inlet temperature to ~380°C, but in this case the maximum inlet concentration of $SO_2$ appears to be even lower (~12.0%) due to a lower thermal stability limit for this catalyst (~620°C). Notably, this relates to the special, rather expensive, types of vanadia catalysts, while the widespread cheaper types are characterized by worse temperature window parameters (from 410–420°C up to 620°C), so in many cases the sulfuric acid reactors are operated with the inlet $SO_2$ concentrations not exceeding 9–10% vol.

As seen from Figure 5.17, the Pt/GFCs, which are able to work at the inlet temperature of ~350°C and have a thermal stability limit not less than 650°C, make it possible to increase significantly (up to ~18.5% vol.) the maximal $SO_2$ concentration in the initial feed gas, compared to vanadia catalysts. It opens the possibility

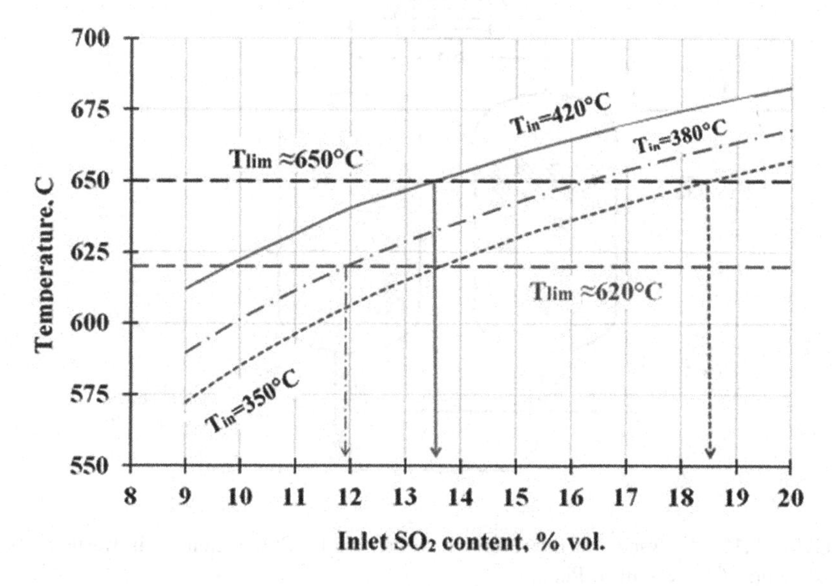


**FIGURE 5.17** Equilibrium adiabatic temperature at the outlet from the first catalytic bed of $SO_2$ oxidation reactor vs. inlet sulfur dioxide concentration for different inlet gas temperatures.

to enhance the production capacity of the existing sulfuric acid plants without any capital investments for new equipment by the simple replacement of the traditional vanadia catalyst in the first reactor bed with Pt/GFC.

### 5.3.2 Reverse-Flow Process With the Additional GFC Beds for Smelter Gas Processing, Which Contains CO and $SO_2$

The reverse-flow process, based on the catalytic oxidation of sulfur dioxide under periodical flow reversal in the catalyst bed, is an efficient technology for the processing of various waste gases of non-ferrous smelters, containing from 1.5–2% to 7–10% vol. $SO_2$ with the production of sulfuric acid.[28] This process provides a stable and energy-saving operation under the significant fluctuations of waste gas parameters (composition, temperature and flow rate) in time, typical for waste gases in non-ferrous metallurgy.[28–30]

Brand new smelter techniques[31] with greater efficiency sometimes face a problem when carbon monoxide occurs in waste gases in considerable amounts (up to 1–2% vol. CO). CO oxidation can produce the overheating of the catalyst bed, causing the decrease of $SO_2$ equilibrium conversion, and also a significant disturbance of the reactor heat regimes, especially when the CO concentration in the waste gases varies widely (e.g., 0.4–2.0%). Also, CO can provoke the deactivation of the vanadia catalyst.[32]

The transformed GFC-based reverse-flow reactor design was developed[33–35] to overcome the negative CO influence. Figure 5.18 presents the flowsheet of the

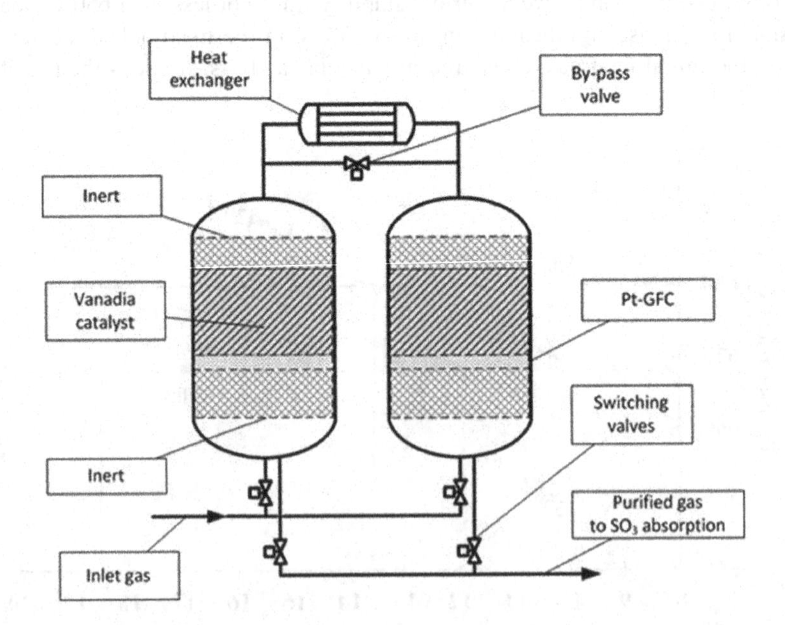


**FIGURE 5.18** Flowsheet of the reverse-flow process for $SO_2$ abatement in waste gases of Mednogorsk Copper-Sulfur Plant.

*Source:* Reprinted with permission from Zagoruiko et al. 2014.[33]

changed process. Pt/GFC beds are made between the beds of inert heat-regenerating material and usual granular vanadia-based catalysts. The primary function of the Pt/GFC inserts is to oxidize CO prior to reaching the vanadia catalyst bed.

The process was simulated using the two-phase one-dimensional mathematical model of the adiabatic plug flow reactor, accounting for the reactions of $SO_2$ and CO oxidation, interphase heat and mass transfer, as well as internal diffusion limitations in the vanadia catalyst pellets.[33]

Kinetic equations, obtained earlier, were used to describe the reaction rates at Pt/GFC for the oxidation of sulfur dioxide:[23]

$$W_1 = k_{01} exp\left(-\frac{E_1}{RT}\right) C_{SO_2} C_{O_2} \left(1 - \frac{C_{SO_3}}{C_{SO_2} C_{O_2}^{\frac{1}{2}} K_p(T)}\right) \tag{5.1}$$

and oxidation of CO:[33]

$$W_2 = k_{02} exp(-E_2 / RT) C_{CO} C_{O_2} \tag{5.2}$$

where $W_1$ and $W_2$ are the $SO_2$ and CO oxidation rates, respectively; $k_{01}$ and $k_{02}$ are the pre-exponents of the kinetic constants; $E_1$ and $E_2$ are the activation energies; $C_i$ are the volume concentrations of the reactants; $K_p(T)$ is the equilibrium constant for the $SO_2$ oxidation reaction; $R$ is the universal gas constant; and $T$ is the temperature.

Rate of the sulfur dioxide oxidation at vanadia catalyst was described by the classic Boreskov-Ivanov equation using the kinetic parameters for the commercial catalyst IC-1–6:[36]

$$W_3 = k_{03} exp(-E_3 / RT) \frac{C_{SO_2} C_{O_2}}{1 + k_2 C_{SO_2}} \left(1 - \frac{C_{SO_3}}{C_{SO_2} C_{O_2}^{0.5} K_p(\theta)}\right) \tag{5.3}$$

The following equation[37] was used for the rate of CO oxidation at vanadia catalyst in the presence of $SO_2$:

$$W_4 = \frac{k_{CO} k_O C_{CO} C_{O_2}}{k_{CO} C_{CO} + k_O C_{O_2}} \tag{5.4}$$

Every simulation with each set of the initial data was performed using the original author's software until the achievement of the established cyclical reverse-flow regime, when process parameters during each cycle were accurately repeated during each next cycle.

The process was simulated in two variants: in its existing configuration and with the additional Pt-GFC beds for CO oxidation, inserted between the vanadia catalyst and inlet inert beds (Figure 5.19).

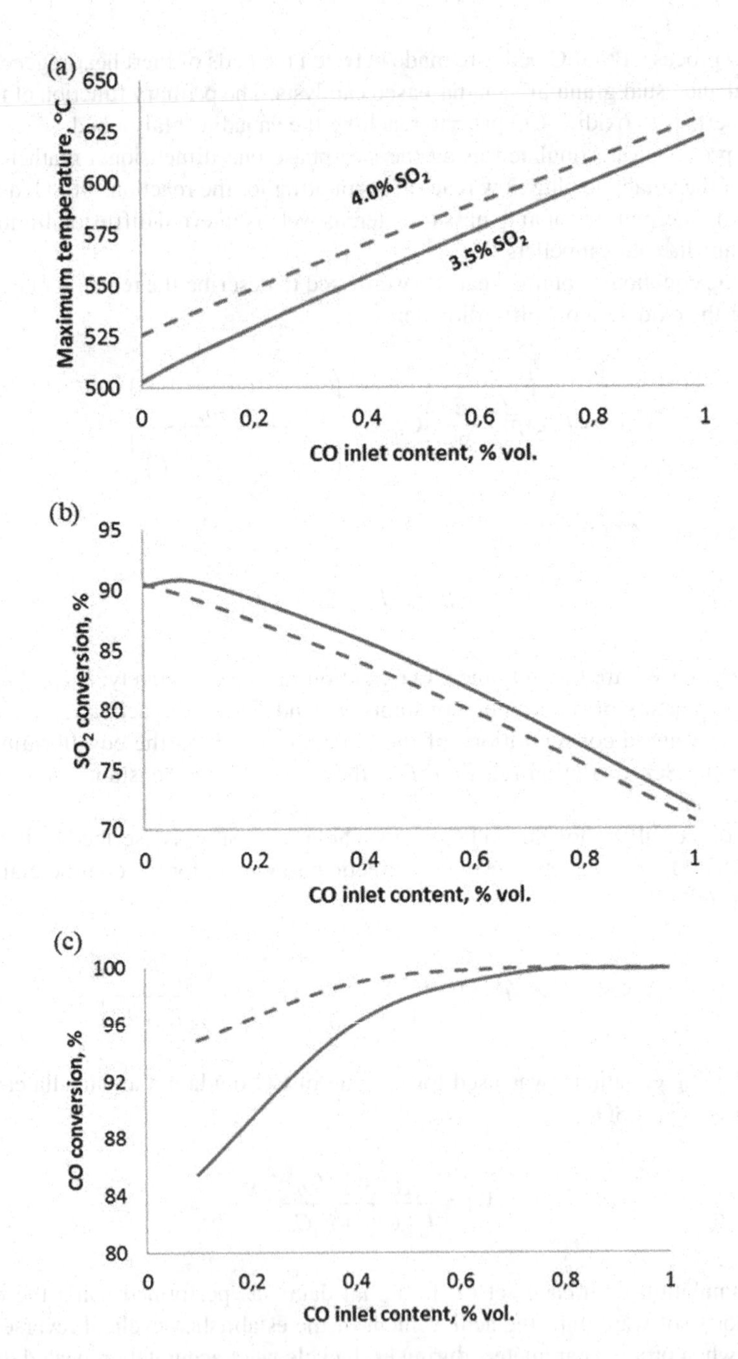


**FIGURE 5.19** Calculated dependence of maximum catalyst temperature (a), $SO_2$ (b), and CO (c) conversions on the inlet CO concentration in the established reverse-flow regime. Inlet gas composition (%vol.): $SO_2$ 3.5% (solid line) or 4.0% (dashed line), $SO_3$ 0.04%, $O_2$ 8%, balance—$N_2$, $T_{in}$ = 50°C, $T_h$ = 400°C, cycle duration = 15 minutes.

*Source:* Reprinted with permission from Zagoruiko et al. 2014.[33]

Figure 5.19 shows the simulated influence of CO content in the inlet gas under the parameters of the conventional reverse flow process (without GFC beds). It is seen that the presence of CO decreases the $SO_2$ conversion significantly, due to temperature rise and corresponding deterioration of the equilibrium conditions.

Moreover, it is seen that the increase of CO concentration from 0 to 1% vol. produces much more negative influence on the process performance than the increase of $SO_2$ inlet concentration by 1% vol. It is caused by a higher heat effect and corresponding adiabatic heat rise for CO oxidation reaction (~100°C per 1% vol. CO compared to ~30°C per 1% $SO_2$). In addition, although the sulfur dioxide oxidation reaction is reversible and temperature rise may lead to a lower equilibrium conversion, partially limiting this heating, the carbon monoxide oxidation is irreversible, creating positive feedback between CO conversion and further temperature rise, which leads to process destabilization in the presence of CO.

Figure 5.20 shows the temperature profiles along the total catalyst bed length averaged in time during the reverse-flow cycle in the established cyclical unsteady-state regime.

In the basic process just described, CO appearance in gas leads to the temperature increase, and it relates both to maximal and average outlet gas temperature. The average outlet temperature in this case is increased by the value of CO oxidation adiabatic heat rise, in strict compliance with the energy conservation law.

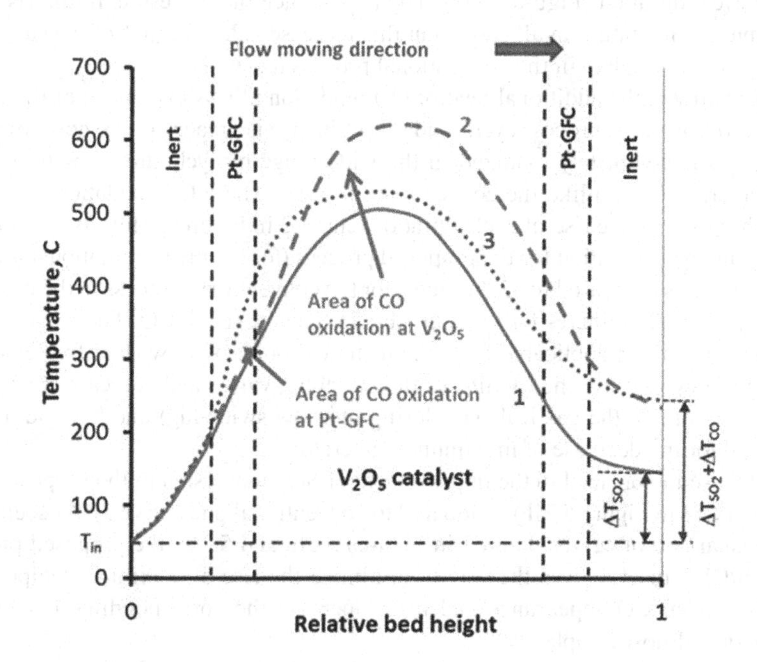


**FIGURE 5.20** Axial temperature profiles averaged per cycle duration in the established reverse-flow regimes for the different cases of the reverse flow processes: (1) existing process in absence of CO, (2) existing process in presence of 0.4% vol. CO, (3) new process with the additional beds of CO oxidation catalyst in presence of 0.4% vol. CO.

*Source:* Reprinted with permission from Zagoruiko et al. 2014.[33]

Simultaneously, the maximum temperature depends not only on the rise of reaction heat emulation, but also on the complicated interplay of many factors, connected with kinetics of reactions as well as with heat and mass transfer processes. If CO starts to oxidize at temperatures above the initial temperature for $SO_2$ oxidation, as happens with the vanadia catalyst, then the CO presence leads to a significant rise of maximum temperature exactly in the center of the catalyst bed, where the main part of CO oxidation occurs. Correspondingly, this temperature rise has a negative influence on the equilibrium conversion of $SO_2$, which is defined by maximum temperature to a significant extent.

In case of application of Pt/GFC inserts, the CO oxidation occurs in these inserts; this oxidation starts at lower temperature than that for $SO_2$ at vanadia catalyst (i.e. below ~400°C). In this situation, oxidation of carbon monoxide, though leading to a higher outlet temperature, produces much less influence on maximum temperature, which helps to avoid the decrease of $SO_2$ in the presence of CO.

Figure 5.21 demonstrates the comparative simulation results for RFR arrangement types: the existing process, when there is no CO in the waste gases, is marked by curve 1; while the curve 2 corresponds to the same process but in case of CO presence; and finally, the suggested process with GFC inserts in the presence of CO is marked by curve 3.

Obviously, in all the cases the maximum temperature decreases with the increase of the cycle duration (Figure 5.21a). The appearance of CO results in the rising of maximum temperature in all cases, but this increase is less in case of the suggested process (curve 3), than in the conventional process (curve 2).

Interestingly, the additional heat of CO oxidation allows expansion of the maximum duration of the process cycle, and this influence is greater for the novel process, where CO is completely oxidized in the wide range of cycle durations (see curve 3 in Figure 5.21c), unlike the conventional process where CO oxidation efficiency falls sharply with the rise of cycle duration (curve 2 in Figure 5.21c). As a result, the maximum cycle duration for the proposed process (for the given conditions) is equal to ~30 minutes compared to ~25 minutes for the conventional process in the presence of CO and to ~22 minutes for same process in the absence of CO. The longer cycles are generally more applicable from the technical point of view (less frequent flow reversals may improve the lifetime of the switching valves and decrease the conversion losses due to the gas leakages during the flow switching) and have additional potential for the decrease of maximum temperature.

All these factors lead to the improvement of $SO_2$ conversion in the proposed process (curve 3 in Figure 5.21b), compared to conventional one (curve 2). As seen from the comparison of curves 3 and 1 in Figures 5.21a and 5.21b, the proposed process with Pt/GFC inserts gives the way to minimize the rise of maximum temperature resulting from CO appearance and to compensate the corresponding loss of $SO_2$ conversion almost completely.

Though CO oxidation was considered here mostly in terms of the impact on $SO_2$ oxidation efficiency, CO abatement in smelter waste gases is an important environmental task itself. In this context, the complete CO conversion (Figure 5.21c) in the wide range of process parameters is an additional advantage of the proposed technology.

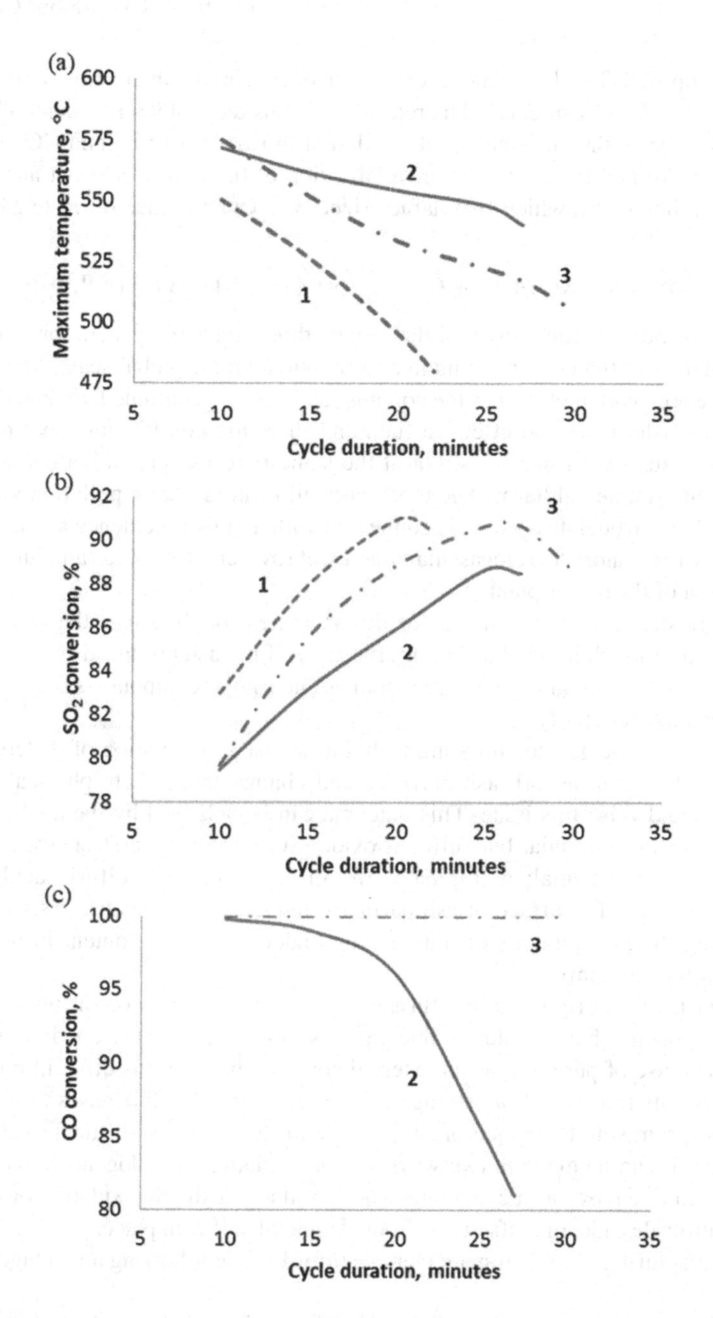


**FIGURE 5.21** Calculated dependence of maximum catalyst temperature (a), $SO_2$ (b), and CO (c) conversions on the cycle duration in the established reverse flow regime for the different process types: (1) existing process in absence of CO, (2) existing process in presence of 0.4% vol. CO, (3) proposed process with Pt-GFC inserts (structured cartridges, 0.2 m height) in presence of 0.4% vol. CO. Inlet gas composition (%vol.): $SO_2$ 4.0%, $SO_3$ 0.04%, $O_2$ 8%, balance—$N_2$, $T_{in}$ = 50°C, $T_h$ = 400°C.

*Source:* Reprinted with permission from Zagoruiko et al. 2014.[33]

The proposed GFC-based modification can be potentially interesting for the application in non-ferrous smelters. The relevance of this technology is provided by both the tightening of the environmental regulation in respect of $SO_2$ and CO emission to atmosphere and the active commercialization of the highly efficient autogenous melting technologies, which are characterized with CO presence in waste gases.

### 5.3.3 Conditioning of Flue Gases from Coal-Fired Power Plants

Regardless the dynamic growth of the new methods in energy generation, coal-fired power plants continue to be the main power source on the global scale. Most likely, the active application of coal as the combustible fuel will continue for some decades. Coal is an affordable and effective fuel, and there are considerable coal reserves, which will suffice for centuries, but at the same time its application causes a significant environmental harm. The most harmful is atmospheric pollution with a fly ash and dust particulates, which is connected with a higher incidence and death rate caused by respiratory, cardiovascular, and cerebrovascular diseases and lung cancer in the area of the power plant.[38]

The most effective instrument for the abatement of fly ash is the electrostatic precipitator, though it may have poor efficiency, if the ash produced by combustion of some coal types, containing high amount of non-polar components like silica, has a high electric resistivity.

The use of flue gas conditioning techniques based on addition of different substances, which can adsorb ash particles and change their electrophysical properties,[39,40] could solve this issue. This outcome can be achieved by the application of water vapor or ammonia, but sulfur trioxide ($SO_3$)[41–43] has the strongest and most comprehensive potential: it originates the microdroplets of sulfuric acid, which are adsorbed on the surface of ash particles, reducing their electric resistivity and enhancing the precipitator's operation even under lower $SO_3$ content in flue gases (e.g., few tens of ppm).

Sulfur trioxide originates in a furnace during the oxidation of sulfur-containing coal components, but its value in flue gases is insufficient for the conditioning purposes. The use of pure $SO_3$ as an external conditioning agent is difficult due to the issues with its transportation, storage and injection into the flue gases flow. Hence, the most promising technique of $SO_3$ supply to flue gases is sulfur dioxide onsite oxidation. It can be made in two ways: by the oxidation of endogenous $SO_2$, which contains in flue gases as a coal combustion product, or the by oxidation of the synthetic sulfur dioxide, manufactured from elemental sulfur in place.

Manufacturing of $SO_3$ from elemental sulfur has the following advantages:

- Complete independence of the process from the flow rate, composition, and temperature of the flue gases.
- Possibility to control $SO_3$ concentration in the flue gases in a wide range.
- Complete absence of dust in the gas flow.
- Possibility to process gases with a high $SO_2$ concentration, providing low gas flow rate and low catalyst loading.
- Absence of limitations for the pressure drop in the catalyst bed.

The calculated characteristics of such process for the conditioning of standard coal-fired boiler TPP-210A with the steam production capacity of 95 tons per hour and electric power 300 MW, which is widely used across the Russian Federation and neighboring countries, are presented in Table 5.5. The flow rate of the flue gases in the referred boiler is 1.2 mln. st.m$^3$/h. According to the information of All-Russian Thermal Physics Institution, the optimal concentration of $SO_3$ in flue gases depends on the type and properties of the used coal, in particular, in case of Kuznetsk coal (South Siberia, Russia) it is equal to 10 ppmv, while for the Ekibastuz coal (Northern Kazakhstan) it reaches 70 ppmv. Oxidation of endogenous sulfur in case of both coal types gives $SO_2$ content in flue gases of 600–900 mg/m$^3$ (~200–300 ppmv).

There are some misgivings about that the increased discharge of $SO_2$ to the atmosphere with flue gases can be caused by the insufficient conversion of sulfur dioxide in the reactor. Nevertheless, this increase should not be higher than 2–3 ppmv, which is not an ultimate value for the generic background content of $SO_2$ in flue gases of 200–300 ppmv.

The provisional selection of catalyst for the process involved the consideration of common vanadium and carbon catalysts and the brand-new catalyst on the base of glass fiber-supports.

The carbon catalysts ensure a perfect low-temperature activity, but this characteristic is not major for the side-stream process, as far as the adequate carbons had a high price, they were excluded. The vanadia catalysts with a reasonable price and extensive experience of the industrial application in $SO_2$ oxidation for more than

**TABLE 5.5**
**Technological Parameters of the Conditioning Process of $SO_3$ Manufacturing from Elemental Sulfur[23]**

| Parameter | Meas. unit | Value | |
|---|---|---|---|
| | | Kuznetsk coals | Ekibastuz coals |
| Gas flow rate | st.m$^3$/hour | 155 | 1085 |
| $SO_2$ conversion in the reactor | % | 78 | |
| Production capacity in respect to $SO_3$ | st.m$^3$/hour | 12 | 84 |
| $SO_3$ content in flue gases | vppm | 10 | 70 |
| Gas temperature at the reactor inlet | °C | 350 | |
| Gas temperature at the reactor outlet | °C | 580 | |
| Catalyst loading | l/kg | 50/6 | 350/40 |
| Catalyst loading cost | USD | 180 | 1200 |
| Reactor pressure drop | mm WC | Not higher than 250 | |
| Energy consumption of the air pump | KW | 0.2 | 1.4 |
| Liquid sulfur consumption | kg/hour | 22 | 154 |
| Annual sulfur consumption | tons | 176 | 1232 |

*Source:* Reprinted with permission from Zagoruiko et al. 2010.[26]

70 years[36] can be used in this process[40] with a good result. Notwithstanding, the catalytic reactor for the specified side-stream process (in comparison with the conventional industrial $SO_2$ oxidation converter) would be distinguished with a relatively small volume, causing inevitable increase of relative heat losses into the environment and thus creating the temperature non-uniformities and cold areas in the catalyst beds. In these conditions vanadia catalysts can be subjected to a high deactivation; hence, the elaboration of the catalysts with the advanced low-temperature activity and operation stability is a critical task for the specified $SO_2$ oxidation process.

In the context of the given study, special attention was paid to Pt-containing GFCs, demonstrating a high catalytic activity in $SO_2$ oxidation reaction (see Chapter 2.2) and high catalyst resistance to poisoning and deactivation in the aggressive reaction medium.[26] It is important that Pt/GFC demonstrates higher activity in the area of relatively low temperatures (below 400°C) and much lower ignition temperature in comparison with the vanadia catalysts. At the same time, the high-temperature activity of GFCs was found to be lower than that for the conventional vanadium catalyst; therefore, it was proposed to use the combined system, including both catalysts, which was expected to provide an efficient operation in the whole range of the process operation temperatures.

The proposed process approach was verified at the pilot-scale unit at Byisk Oleum Plant (Byisk, Altay region, Russia). This plant was installed separately from the main technological pipeline with the feed of $SO_2$-containing gas in the production of sulfuric acid. This gas, containing ~ 8% vol. $SO_2$, is produced by the combustion of elemental sulfur in air, thus exactly reproducing the estimated conditions for the operation of the proposed conditioning technology. The maximum total gas flow supply to the pilot reactor was as high as 120–150 st.m$^3$/hour, being commensurable with the scale of the conditioning installation for the mentioned 300 MW coal-fired boilers. The scheme and the appearance of the pilot reactor are presented in Figure 5.22.

The reactor represents a vertical cylindrical vessel with two conic lids made of high-quality stainless steel. Catalysts were placed to the reactor in two cartridges (6 and 7), one above the other, with the reaction gas moving in the downward direction along the vertical passages formed between GFC layers. Two sockets (8 and 9) with the internal diameter of 10 mm were situated above and under the catalyst bed for sampling of inlet and outlet gas mixture respectively. The analysis of $SO_2$ concentration in the gas samples was provided by gas chromatography; $SO_3$ concentration in the outlet gases was calculated assuming that it is equal to the difference between $SO_2$ inlet and outlet concentrations. U-type manometer (17) was used for the measurement of the pressure drop in the reactor. The reactor and pipelines were heat-insulated by mineral wool.

The structure of the combined cartridge is shown in Figure 5.23. The GFC clothes were twisted in a roll with the layers of the structuring plain metal mesh and bed of the ring-shaped vanadia catalyst pellets, strung on a stainless-steel wire.

Both cartridges had a height of 200 mm; the diameter of the larger cartridge was 500 mm, the small one, 350 mm. Pt/GFC loading was 3.5 and 2.0 kg respectively. The experiments involved various inlet temperatures (up to 500°C) and gas flow rates (55–120 st.m$^3$/h). Inlet content of $SO_2$ was fluctuating in the range of 7–9% vol.

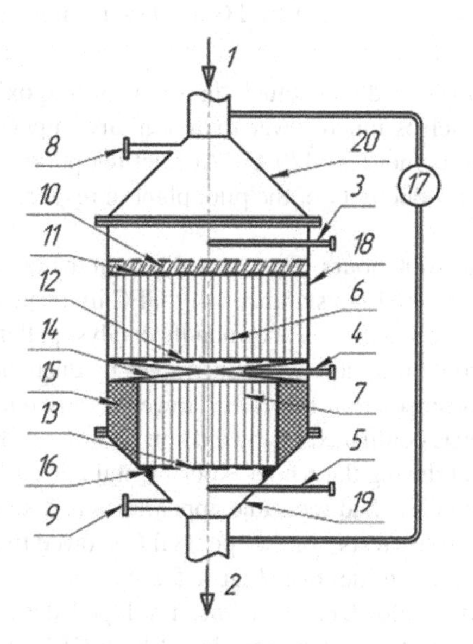

**FIGURE 5.22** The scheme of the pilot reactor in Biysk (left): (1, 2) inlet and outlet gas flows, (3–5) thermocouples, (6, 7) catalytic cartridges, (8, 9) gas sampling sockets for inlet and outlet gas, (10) ceramic packing for the improved flow distribution, (11–13) metal gauzes, (14) supporting cross-piece, (15) mineral wool insulation, (16) supporting arms, (17) U-type manometer, (18) reactor cowling, (19, 20) reactor vessel lids. The appearance of the pilot reactor is given on the right.

*Source:* Reprinted with permission from Zagoruiko et al. 2010.[26]

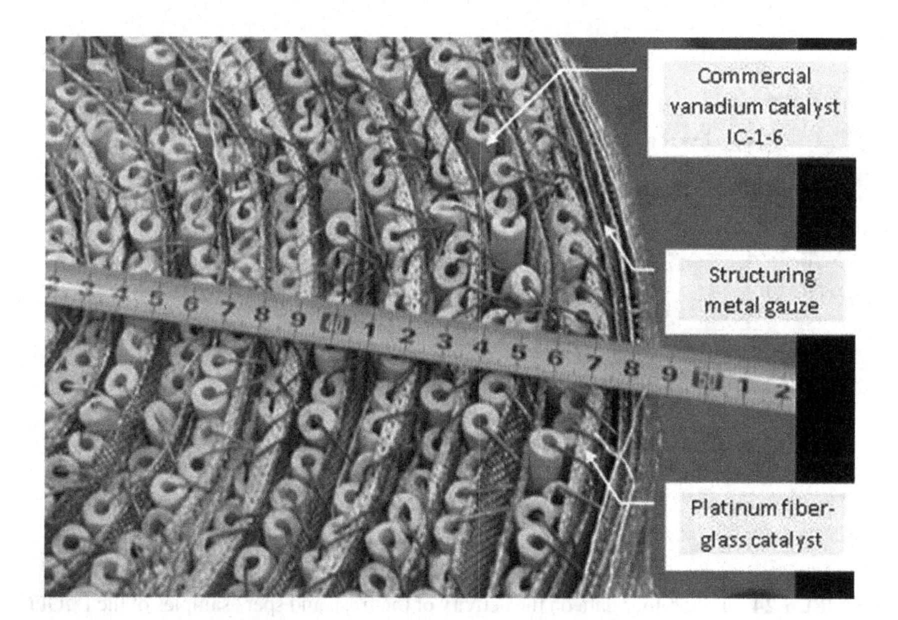


**FIGURE 5.23** Structure of the combined catalytic cartridge.

*Source:* Reprinted with permission from Zagoruiko et al. 2010.[26]

In the course of experiments, it was discovered that actual "ignition" of $SO_2$ oxidation reaction occurs at 380–385°C, which is much lower than similar temperature for vanadia catalyst in these conditions (not less 420°C). At inlet temperature ~420°C and gas flow rate 100 st.m$^3$/h the productivity of the pilot plant in respect of $SO_3$ was equal to 3 st.m$^3$/h.[26]

The tests were performed for more than 1000 hours. The temperature in the reactor was constantly increasing, thus testifying the high stability of Pt/GFC in the process conditions. After the tests, the partial destruction of the vanadia catalyst pellets was discovered, as well as insignificant contamination of the cartridges by dust and pieces of mineral wool, used as a thermo-insulation material. However, this phenomenon was not of critical character. It was confirmed by the data on pressure drop of the pilot plant, which remained constant during the whole experimental period.[23]

After unloading from the cartridge, the GFC had the same appearance and kept its flexibility completely. The additional activity tests[27] of the spent GFC showed that its activity appeared to be even higher than the initial one (Figure 5.24).

We may conclude that according to the pilot trial data, the developed design provided an efficient combination of the advantageous properties of both GFC and granular vanadia catalyst. The produced combined cartridges demonstrated a stable performance with no signs of GFC deactivation and without increase of pressure drop during long-term (more than 1000 hours) continuous operation in the medium of $SO_2$-containing gases.

The disadvantage of the proposed technology is a consumption of side feedstock (liquid sulfur), though according to our estimations the overall sulfur consumption appears to be appropriate (~ 20 kg/hour for a standard 300 MW boiler fed with typical Russian coals, such as Kuznetsk one).

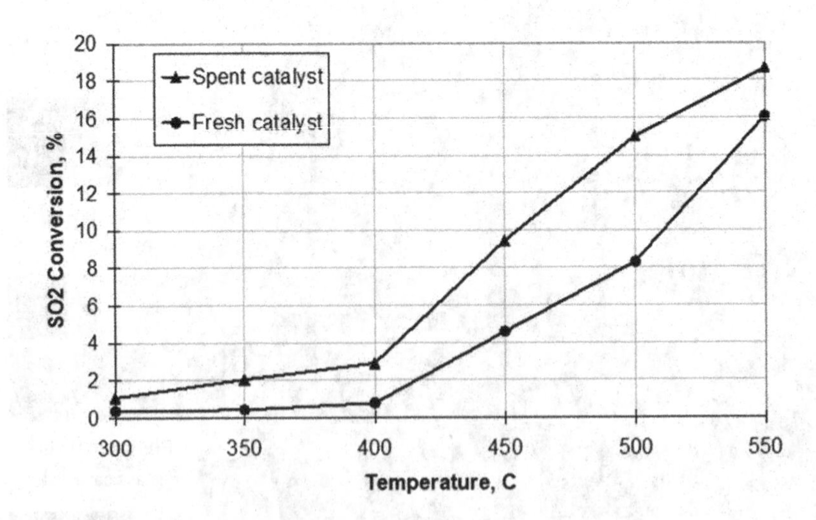


**FIGURE 5.24** Laboratory data on the activity of the fresh and spent samples of the Pt/GFC. Catalyst loading in each experiment was equal to 0.17 g; gas flow rate 51 mL/min; reaction gas composition: $SO_2$ 1.67% vol., O2 3.33% vol., balance—helium.

Nevertheless, the development of a novel conditioning technology, based on oxidation of endogenous $SO_2$, naturally presented in flue gases is an actual task as well. It seems possible to solve this problem by the application of structured GFC cartridges, placed directly to the flue gas duct.[44] The corrugation free (see Figure 3.11) cartridges and lemniscate bed (Figure 3.16), both characterized with an extra-low pressure drop, look advantageous for this application.

Due to high activity of Pt/GFC in oxidation of the organic compounds, its application can be useful for the settlement of two related environmental problems, which currently attracts a lot of attention in respect to flue gases from coal-fired power plants: incineration of traces of polyaromatic compounds, formed during coal combustion, and oxidation of mercury-containing organic substances.

## REFERENCES

1. Zagoruiko A.N., Lopatin S.A., Bal'zhinimaev B.S. et al. 2010. The process for catalytic incineration of waste gas on IC-12-S102 platinum glass fiber catalyst. *Catalysis in Industry.* 2: 113–17.
2. Zagoruiko A.N., Bal'zhinimaev B.S., Lopatin S.A. et al. 2010. Commercial process for incineration of VOC in waste gases on the base fiber-glass catalyst. Proceedings of XIX International Conference on Chemical Reactors CHEMREACTOR-19 (Vienna, Austria, September 5–9): 586–7.
3. Zagoruiko A.N., Bal'zhinimaev B.S., Arendarskii D.A. et al. 2004. Russian Patent No.2231653. Device for purification of exhaust gases from internal combustion engines.
4. Arendarskiy D.A., Zagoruyko A.N., Bal'zhinimaev B.S. 2005. Glass-fibre catalysts to clear diesel engine exhausts. *Chemistry for Sustainable Development.* 13(6): 731–5.
5. Zagoruiko A.N., Lopatin S.A., Klenov O.P. 2013. Russian Patent for Utility Model No.125094. Catalytic system for performance of heterogeneous reactions.
6. Zagoruiko A.N., Lopatin S.A., Klenov O.P. 2013. Russian Patent for Utility Model No.124888. Reactor for performance of heterogeneous catalytic process.
7. Serbinenko V.V., Zagoruiko A.N., Lopatin S.A. et al. 2013. Russian Patent for Utility Model No.124925. Catalytic system for purification of diesel exhausts.
8. Makhov S.F., Ushakov S.N., Baskakov V.S. et al. 2013. Russian Patent for Utility Model No.124924. Catalytic system for purification and cooling of diesel exhausts.
9. Lopatin S.A., Tsyrul'nikov P.G., Kotolevich Y.S. et al. 2015. Structured woven glass-fiber IC-12-S111 catalyst for the deep oxidation of organic compounds. *Catalysis in Industry.* 7(4): 329–34.
10. Baskakov V.S., Serbinenko V.V., Mishchenko P.A. et al. 2012. Reactor for purification and cooling the exhaust gases from the stationary diesel engines. Proceedings of the XX International Conference on Chemical Reactors CHEMREACTOR-20 (Luxemburg, December 3–7): 173–4.
11. Lopatin S., Chub O., Yazykov N. et al. 2014. Structured cartridges with reinforced fiber-glass catalyst for fuel combustion in the fluidized beds of the inert heat-transfer particles. Proceedings of the XXI International Conference on Chemical Reactors "CHEMREACTOR-21" (Delft, The Netherlands, September 22–25): 272–3.
12. Parmon V.N., Simonov A.D., Sadykov V.A. et al. 2015. Catalytic combustion: Achievements and problems. *Combust Explos Shock Waves.* 51: 143–50.
13. Ismagilov Z.R. 1992. Catalysis and environment. *Appl Catal A-Gen.* 84: N14–N17.
14. Zagoruiko A.N., Lopatin S.A. 2018. Russian Patent No.2674231. Method for combustion of gaseous fuels and device for its performance.

15. Lopatin S.A., Mikenin P.E., Elyshev A.V. et al. 2019. Catalytic device on the base of glass-fiber catalyst for environmentally safe combustion of fuels and utilization of toxic wastes. *Chem Eng J.*, accepted to publication. DOI: 10.1016/j.cej.2019.05.056
16. Zagoruiko A.N., Lopatin S.A., Mikenin P.E. et al. 2017. Novel structured catalytic systems—cartridges on the base of fibrous catalysts. *Chem Eng Proc: Proc Int.* 122: 460–72.
17. Catalytic air heater, http://en.catalysis.ru/block/index.php?ID=27&SECTION_ID=1714
18. Shikina N.V., Yashnik S.A., Gavrilova A.A. et al. 2018. Formation of active structures in monolith copper—manganese oxide catalysts for air-heating devices. *Kinet Catal.* 59: 532.
19. Paukshtis E.A., Simonova L.G., Zagoruiko A.N. et al. 2010. Oxidative destruction of chlorinated hydrocarbons on Pt-containing fiber-glass catalysts. *Chemosphere.* 79(2): 199–204.
20. Bal'zhinimaev B.S., Paukshtis E.A., Simonova L.G. et al. 2004. Oxidative destruction of chloroorganic compounds at glass-fiber catalysts. *Catalysis in Industry.* 5: 21–7.
21. Bal'zhinimaev B.S., Paukshtis E.A., Vanag S.V. et al. 2010. Glass-fiber catalysts: Novel oxidation catalysts, catalytic technologies for environmental protection. *Catal Today.* 151(1–2): 195–9.
22. Zagoruiko A.N., Glotov V.D., Lopatin S.A. et al. 2016. Investigation of the internal structure, fluid flow dynamics and mass transfer in the multi-layered packing of glass-fiber catalyst in the pilot reactor for sulfur dioxide oxidation. *Science Bulletin of the Novosibirsk State Technical University.* 3(64): 161–77.
23. Vanag S.V. 2012. Processes for oxidation of $SO_2$ into $SO_3$ using glass-fiber catalysts and their equipment arrangement. PhD Thesis. Novosibirsk: Boreskov Institute of Catalysis.
24. Vanag S.V., Zagoruiko A.N., Zykov A.M. 2012. Chemisorption and oxidation of $SO_2$ at Pt-containing fiber-glass catalysts. Proceedings of IX International Conference on Mechanisms of Catalytic Reactions (St. Petersburg, October 22–25): 272.
25. Zagoruiko A., Vanag S., Bal'zhinimaev B. et al. 2009. Catalytic flue gas conditioning in electrostatic precipitators of coal-fired power plants. *Chem Eng J.* 154: 325–32.
26. Zagoruiko A., Bal'zhinimaev B., Vanag S. et al. 2010. Novel catalytic process for flue gas conditioning in electrostatic precipitators of coal-fired power plants. *J Air Waste Manag Assoc.* 60: 1002–8.
27. Vanag S.V., Paukshtis E.A., Zagoruiko A.N. 2015. Properties of platinum-containing glass-fiber catalysts in the $SO_2$ oxidation reaction. *React Kinet Mech Cat.* 116: 147–58.
28. Matros Y.S. 1985. *Unsteady processes in catalytic reactors.* Amsterdam: Elsevier Science Publishers.
29. Matros Y.S., Bunimovich G.A. 1996. Reverse-flow operation in fixed bed catalytic reactors. *Catal Rev.* 38(1): 1–68.
30. Zagoruiko A.N. 2012. The reverse-flow operation of catalytic reactors: History and prospects. *Current Topics in Catalysis.* 10: 113–29.
31. Moskalyk R.R., Alfantazi A.M. 2003. Review of copper pyrometallurgical practice: Today and tomorrow. *Miner. Eng.* 16: 893–919.
32. Wen-De Xiao, Hui Wang and Wei-KangYuan. 1999. Practical studies of the commercial flow-reversed $SO_2$ converter. *Chem. Eng. Sci.* 54: 4645–52.
33. Zagoruiko A.N., Vanag S.V. 2014. Reverse-flow reactor concept for combined $SO_2$ and CO oxidation in smelter off-gases. *Chem Eng J.* 238: 86–92.
34. Zagoruiko A.N., Vanag S.V. 2012. Reverse-flow reactor concept for combined $SO_2$ and CO oxidation in smelter off-gases. Proceedings of the XX International Conference on Chemical Reactors CHEMREACTOR-20 (Luxemburg, December 3–7): 70–1.

35. Vanag S.V., Zagoruiko A.N. 2011. Process of combined oxidation of CO and $SO_2$ in waste gases of non-ferrous smelters at platinum glass-fiber catalyst. Proceedings of International Conference "Nanostructured Catalysts and Catalytic Processes for the Innovative Energetics and Sustainable Development" (Novosibirsk, Russia, June 6–10): 72.
36. Amelin A.G. 1983. *Technology of sulfuric acid.* Moscow: Chimia.
37. Vorlow S., Wainwright M.S., Trimm D.L. 1985. The catalytic activity and selectivity of supported vanadia catalysts doped with alkali metal sulphates. II. The role of sulfur in determining activity and selectivity of carbon monoxide and benzene oxidation. *Applied Catalysis.* 17: 103–14.
38. Kravchenko J., Kim Lyerly H. 2018. The impact of coal-powered electrical plants and coal ash impoundments on the health of residential communities. *N C Med J.* 79: 289–300.
39. Dalmon J., Tidy D. 1972. A comparison of chemical additives as aids to the electrostatic precipitation of fly-ash. *Atmos Environ* (1967). 6(10): 721–2, IN1–IN2, 723–34.
40. Shanthakumar S., Singh D.N., Phadke R.C. 2008. Flue gas conditioning for reducing suspended particulate matter from thermal power stations. *Prog Energ Combust.* 34(6): 685–95.
41. Navarrete B., Alonso-Fariñas B., Lupiõn M. et al. 2015. Effect of flue gas conditioning on the cohesive forces in fly ash layers in electrostatic precipitation. *Environ Prog Sustain.* 34(5): 1379–83.
42. Qi L., Yuan Y. 2013. Influence of $SO_3$ in flue gas on electrostatic precipitability of high-alumina coal fly ash from a power plant in china. *Powder Technol.* 245: 163–7.
43. Wu H., Pan D., Zhang R. et al. 2017. Abatement of fine particle emissions from a coal-fired power plant based on the condensation of $SO_3$ and water vapor. *Energ Fuel.* 31(3): 3219–26.
44. Zagoruiko A.N., Mikenin P.E., Lopatin S.A. et al. 2018. $SO_2$ oxidation in structured catalytic cartridges with glass-fiber catalyst for conditioning of flue gases from coal-fired powerplants. Proceedings of XXIII International Conference on Chemical Reactors CHEMREACTOR-23 (Ghent, Belgium, November 04–11): 400–1.

# Conclusions

As shown by the performed studies, a novel catalyst can be created by the means of supporting the various active components at the microfibrous glass-fiber fabric. Glass-fiber catalysts (GFCs) are able to use the noble metals (like Pt or Pd) or transient metal oxides as the active components, thus providing quite a wide range of the possible practical applications. The surface thermal synthesis (STS) method enables synthesis of highly dispersed active sites, characterized by high activity and operation stability in many catalytic reactions. The proposed method is based on the use of inexpensive and accessible materials; it is technologically simple, easily scalable, and does not produce waste.

Moreover, the main advantage of GFCs is their unusual engineering properties. First of all, it relates to their mechanical flexibility, which opens the way for the creation of new types of the structured catalytic beds with the improved heat and mass transfer, low pressure drop, and high resistance to mechanical and thermal shocks. The undertaken research has demonstrated, that structured GFC cartridges excel the conventional catalysts of all known shapes in the combination of their practically important features (apparent activity, specific pressure drop, etc.).

The proposed GFCs have a high potential for the use in the processes of oxidation of CO, deep oxidation of volatile organic compounds, reduction of nitrogen oxides, oxidation of $SO_2$, and selective oxidation of $H_2S$. First examples of their practical application have confirmed their high efficiency and important advantages over the traditional catalysts.

Though much has been already done in this area, still much more needed to be done to provide GFC with their rightful place in the field of modern catalytic technologies.

# Index

## A



## C



## E



## F



## G



## H



## I



## L



## M



## N



## P



## R



## S



## T



## V



## W

# Index

Printed in the United States
by Baker & Taylor Publisher Services